A Photographic Atlas
for the
Anatomy and Physiology Laboratory

THIRD EDITION

Kent M. Van De Graaff

John L. Crawley

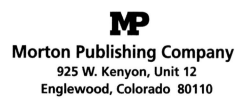

Morton Publishing Company
925 W. Kenyon, Unit 12
Englewood, Colorado 80110

To our wives, Karen and Nancy, and to our children
for their sacrifice, support, and love.

Copyright 1992, 1994, 1996 by Morton Publishing Company

ISBN: 0-89582-313-6

10 9 8 7 6 5 4 3 2 1

Printed in the United States of America

Table of Contents

A Photographic Atlas for the Anatomy and Physiology Laboratory

Preface

Human anatomy is the scientific discipline that investigates the structure of the body and human physiology is the scientific discipline that investigates how body structures function. These subjects may be taught independent of each other in separate courses, or they may be taught together in integrated anatomy and physiology courses. Regardless of whether or not anatomy is taught independently from physiology or if the two disciplines are integrated as a single course, it is necessary for a student to have a conceptualized visualization of body structure and a knowledge of it's basic descriptive anatomical terminology in order to understand how the body functions.

A Photographic Atlas for the Anatomy and Physiology Laboratory is designed for all students taking separate or integrated courses in human anatomy and physiology. This atlas can accompany and will augment any human anatomy, human physiology, or combined human anatomy and physiology textbook. It is designed to be of particular value to students in a laboratory situation and could either accompany a laboratory manual or in certain courses, serve as the laboratory manual. The preparation of *A Photographic Atlas for the Anatomy and Physiology Laboratory* in full color is a valuable improvement of this edition over the previous editions.

Anatomy and physiology are visually oriented sciences. Great care has gone into the preparation of this photographic atlas to provide students with a complete set of photographs for each of the human body systems. Human cadavers have been carefully dissected and photographs taken that clearly depict each of the principal organs from each of the body systems. Cat dissection, fetal pig dissection, and rat dissection are also included for those students who have the opportunity to do similar dissection as part of their laboratory requirement. In addition, photographs of a sheep heart dissection are also included.

A visual balance is achieved in this atlas between the various levels available to observe the structure of the body. Microscopic anatomy is presented by photomicrographs at the light microscope level and electron micrography from scanning and transmission electron microscopy. Carefully selected photographs are used throughout the atlas to provide a balanced perspective of the gross anatomy. At the request of several professors who used previous editions of the atlas, the histology sections have been expanded and improved through procurement of over 150 new photomicrographs. Selected X-rays, CT scans, and MR images depict structures from living persons and thus provide an applied dimension to the atlas. Great care has been taken to construct completely labeled, informative figures that are depicted clearly and accurately. The terminology used in this atlas are those that are approved and recommended by the *Basle Nomina Anatomica (BNA)*.

Acknowledgments

Many professionals assisted in the preparation of this atlas in it's current and previous editions. They share our enthusiasm of it's value for students of anatomy and physiology. We are especially appreciative of William B. Winborn, Ph.D. and Kyle M. Van De Graaff at the University of Texas Health Science Center at San Antonio for their efforts and generosity in providing the choice photomicrographs used in this atlas. The radiographs, CT scans, and MR images have been made possible through the generosity of Gary M. Watts, MD. and the Department of Radiology at Utah Valley Regional Medical Center. Ryan L. Van De Graaff and Kerry Peterson at the University of Utah assisted in dissections of laboratory specimens and in taking photographs. Others who aided in specimen dissections were Nathan A. Jacobson, R. Richard Rasmussen, and Sandra E. Sephton. The student models that we photographed for use in chapters 1 and 7 were particularly helpful. We appreciate the talents of Christopher H. Creek who rendered the line art throughout the book. Many users and reviewers of the previous editions of this atlas provided suggestions for its improvement. We are especially appreciative of Michael J. Shively for his numerous comments and helpful suggestions. The greatest thanks, however, is extended to Brian C. Watts who labored tirelessly in creating the layout of this atlas. We are indebted to Douglas Morton and the personnel at Morton Publishing Company for the opportunity, encouragement, and support to complete this project.

Body Organization

Anatomy is the study of body structures. An example of an anatomical study is learning about the structure of the heart—the chambers, valves, and vessels that serve the heart muscle. *Physiology* is the study of body function. An example of a physiological study is learning what causes the heart muscles to contract—the sequence of blood flow through the heart and what causes blood pressure. The anatomy (structure) and the physiology (function) of any part of the body are always related, or in other words, structure determines function.

Most of the physiological processes within the body act to maintain *homeostasis*. Simply defined, homeostasis is maintaining near consistent internal conditions within the body despite changing conditions in the external environment. For example, one area of your brain acts as a thermostat to keep your body temperature near 37°C (98.6°F). Being too warm causes you to sweat and cool the body, while being cold causes you to shiver and warm the body. Maintaining overall body homeostasis is achieved through many interacting physiological processes involving all levels of body organization, and is absolutely necessary for survival.

Structural and functional levels of organization exist in the body, and each of its parts contributes to the total organism. In the study of human anatomy and physiology, six levels of body organization are generally recognized—the molecular level, the cellular level, the tissue level, the organ level, the systems level, and the organismic level (fig. 1.1).

Cells are microscopic and are the smallest living part of all organisms. *Tissues* are layers of groups of similar cells that perform specific functions. An *organ* is an aggregate of two or more tissues integrated to perform a particular function. The *systems* of the body consist of various body organs that have similar or related functions. All the systems of the body are interrelated and function together constituting the *organism.*

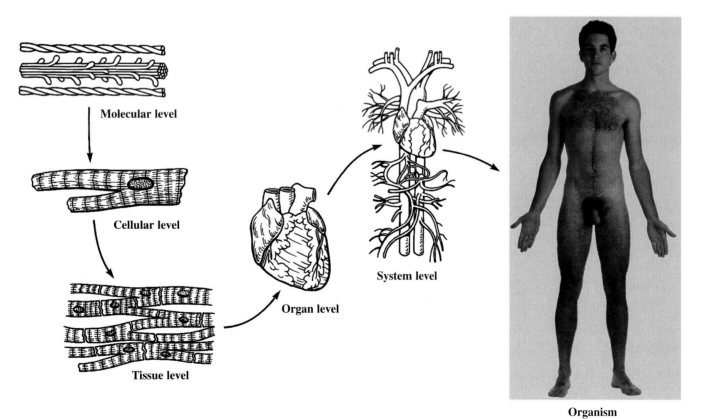

Figure 1.1 The levels of structural organization and complexity within the body.

Table 1.1 Directional Terminology for Describing Human Body Structures

Term	Definition	Example
Superior (cephalic, cranial)	Toward the top of the head	The neck is superior to the thorax.
Inferior (caudal)	Away from the top of the head	The pubic region is inferior to the abdomen.
Anterior (ventral)	Toward the front of the body	The eyes, nose, and mouth are on the anterior side of the body.
Posterior (dorsal)	Toward the back of the body	The spinal cord extends down the posterior side of the body.
Lateral	Toward the side of the body	The arms are on the lateral sides of the body.
Medial	Toward the median plane of the body	The heart is medial to the lungs.
Superficial (external)	Toward the surface of the body	The skin is superficial to the muscles.
Deep (internal)	Away from the surface of the body	The heart is positioned deep within the thoracic cavity.
Parietal	Reference to the body wall of the trunk (thorax and abdomen)	The parietal peritoneum is the membrane lining of the abdominal cavity.
Visceral	Reference to internal organs of trunk	The stomach is covered by a thin membrane called the visceral peritoneum.

Figure 1.2 The anatomical position provides a basis of reference for describing the relationship of one body part to another. In the anatomical position, the person is standing, the feet are parallel, the eyes are directed forward, and the arms are to the sides with the palms turned forward and the fingers are pointed straight down.

Figure 1.3 Major body parts and regions in humans (bipedal vertebrate). (a) An anterior view and (b) a posterior view.

1. Upper extremity
2. Lower extremity
3. Head
4. Neck, anterior aspect
5. Thorax (chest)
6. Abdomen
7. Cubital fossa
8. Pubic region
9. Palmar region (palm)
10. Patellar region (patella)
11. Cervical region
12. Shoulder
13. Axilla (armpit)
14. Brachium (upper arm)
15. Lumbar region
16. Elbow
17. Antebrachium (forearm)
18. Gluteal region (buttock)
19. Dorsum of hand
20. Thigh
21. Popliteal fossa
22. Calf
23. Plantar surface (sole)

Figure 1.4 Directional terminology and superficial structures in a fetal pig (quadrupedal vertebrate).

1. Anus
2. Tail
3. Scrotum
4. Knee
5. Teat
6. Ankle
7. Umbilical cord
8. Hoof
9. Auricle (pinna)
10. External auditory canal
11. Superior palpebra (superior eyelid)
12. Elbow
13. Wrist
14. Naris (nostril)
15. Tongue

Figure 1.5 Directional terminology and superficial structures in a cat (quadrupedal vertebrate).

1. Thigh
2. Tail
3. Auricle (pinna)
4. Superior palpebra (superior eyelid)
5. Bridge of nose
6. Naris (nostril)
7. Vibrissae
8. Brachium
9. Manus (front foot)
10. Claw
11. Antebrachium

Figure 1.6 Planes of reference in a cat.

1. Coronal plane (frontal plane)
2. Transverse plane (cross-sectional plane)
3. Midsagittal plane (median plane)

Figure 1.7 Planes of reference in a human (bipedal vertebrate).

1. Coronal plane (frontal plane)
2. Transverse plane (cross-sectional plane)
3. Midsagittal plane (median plane)

Figure 1.8 MR image of the trunk showing the body cavities and their contents.

1. Thoracic cavity
2. Abdominopelvic cavity
3. Image of heart
4. Image of diaphragm
5. Image of rib
6. Image of lumbar vertebra
7. Image of Ilium

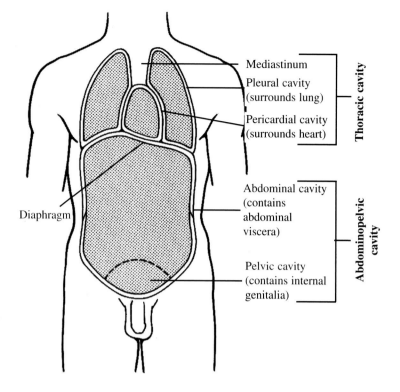

Figure 1.9 An anterior view of the body cavities of the trunk.

Mediastinum
Pleural cavity (surrounds lung)
Pericardial cavity (surrounds heart)

Thoracic cavity

Diaphragm

Abdominal cavity (contains abdominal viscera)

Pelvic cavity (contains internal genitalia)

Abdominopelvic cavity

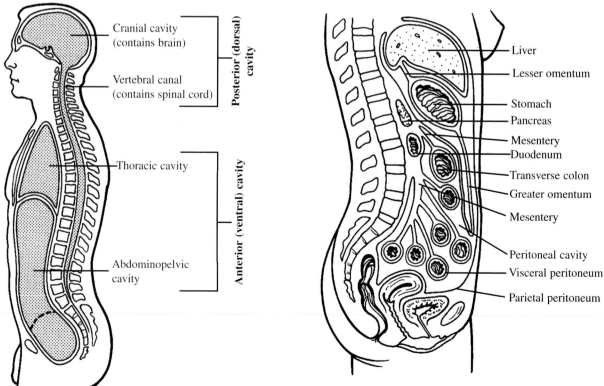

Cranial cavity (contains brain)

Vertebral canal (contains spinal cord)

Posterior (dorsal) cavity

Thoracic cavity

Abdominopelvic cavity

Anterior (ventral) cavity

Figure 1.10 A midsagittal view of the body cavities.

Liver
Lesser omentum
Stomach
Pancreas
Mesentery
Duodenum
Transverse colon
Greater omentum
Mesentery
Peritoneal cavity
Visceral peritoneum
Parietal peritoneum

Figure 1.11 A midsagittal view of the organs of the abdominopelvic cavity and their supporting membranes.

Figure 1.12 Human male.
(a) An anterior view and
(b) a posterior view.

1. Facial region
2. Cranial region
3. Posterior neck
4. Anterior neck
5. Shoulder
6. Thorax
7. Nipple
8. Brachium
9. Elbow
10. Cubital fossa
11. Abdomen
12. Umbilicus (navel)
13. Antebrachium
14. Wrist
15. Hand
16. Natal (gluteal) cleft
17. Fold of buttock
18. External genitalia
19. Thigh
20. Patella
21. Popliteal fossa
22. Leg
23. Ankle
24. Foot

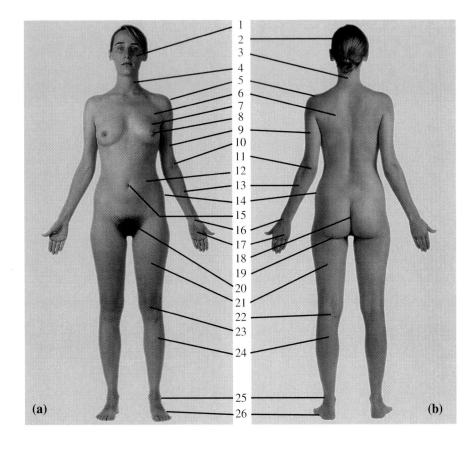

Figure 1.13 Human
female. (a) An anterior view
and (b) a posterior view.

1. Facial region
2. Cranial region
3. Posterior neck
4. Anterior neck
5. Shoulder
6. Thorax
7. Breast
8. Nipple
9. Brachium
10. Cubital fossa
11. Elbow
12. Abdomen
13. Antebrachium
14. Iliac crest
15. Umbilicus (navel)
16. Wrist
17. Hand
18. Natal (gluteal) cleft
19. Fold of buttock
20. Mons pubis
21. Thigh
22. Popliteal fossa
23. Patella
24. Leg
25. Ankle
26. Foot

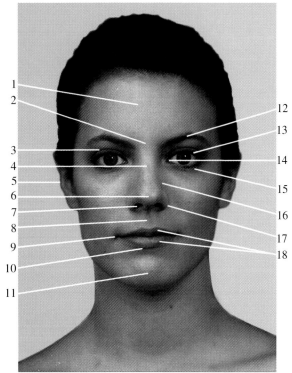

Figure 1.14 An anterior view of the facial region.

1. Forehead
2. Root of nose
3. Superior palpebral sulcus
4. Bridge of nose
5. Auricle (pinna)
6. Apex of nose
7. Nostril
8. Philtrum
9. Corner of mouth
10. Mentolabial sulcus
11. Chin (mentalis)
12. Eyebrow
13. Eyelashes of upper eyelid
14. Lacrimal caruncle
15. Eyelashes of lower eyelid
16. Nasofacial angle
17. Alar nasal sulcus
18. Lips

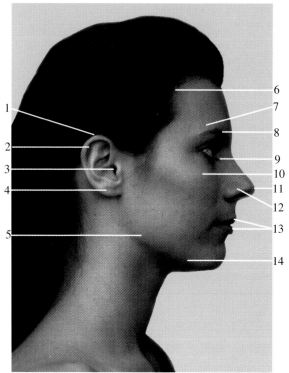

Figure 1.15 A lateral view of the facial region.

1. Helix of auricle
2. Antihelix
3. External auditory canal
4. Earlobe
5. Angle of mandible
6. Hair line
7. Superciliary ridge
8. Eyebrow
9. Eyelashes
10. Zygomatic arch
11. Apex of nose
12. Ala nasi
13. Lips
14. Body of mandible

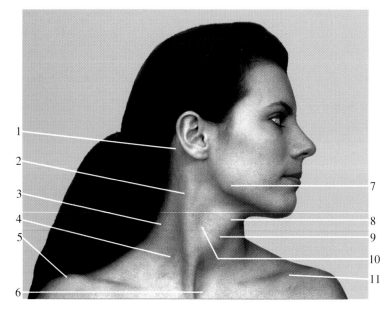

Figure 1.16 An anterolateral view of the neck.

1. Mastoid process
2. Sternocleidomastoid muscle
3. Trapezius muscle
4. Posterior triangle of neck
5. Acromion of scapula
6. Jugular notch
7. Angle of mandible
8. Hyoid bone
9. Thyroid cartilage of larynx
10. Anterior triangle of neck
11. Clavicle

Figure 1.17 An anterior view of the thorax and abdomen.

1. Jugular notch
2. Acromion of scapula
3. Clavipectoral triangle
4. Sternum
5. Nipple
6. Serratus anterior m.
7. Rectus abdominis m.
8. Linea semilunaris
9. External abdominal oblique m.
10. Iliac crest
11. Inguinal ligament
12. Trapezius m.
13. Supraclavicular fossa
14. Clavicle
15. Pectoralis major m.
16. Anterior axillary fold
17. Xiphoid process
18. Linea alba
19. Tendinous inscription
20. Umbilicus

Figure 1.18 An anterolateral view of the thorax, abdomen, and axilla.

1. Triceps brachii m.
2. Biceps brachii m.
3. Sternocleidomastoid m.
4. Deltoid m.
5. Axilla
6. Pectoralis major m.
7. Latissimus dorsi m.
8. Nipple
9. Serratus anterior m.
10. Intercostal m.
11. Linea alba
12. External abdominal oblique m.

Figure 1.19 Right shoulder, axilla, and upper extremity.

1. Olecranon of ulna
2. Sulcus of ulnar nerve
3. Medial epicondyle of humerus
4. Axilla
5. Triceps brachii m.
6. Latissimus dorsi m. (posterior axillary fold)
7. Tendon of flexor carpi radialis longus m.
8. Brachioradialis m.
9. Deltoid m.
10. Biceps brachii m.
11. Pectoralis major m. (anterior axillary fold)

Figure 1.20 A posterior view of the thorax.

1. Deltoid m.
2. Trapezius m.
3. Infraspinatus m.
4. Triangle of auscultation
5. Inferior angle of scapula
6. Latissimus dorsi m.
7. Median furrow over vertebral column
8. Erector spinae m.

Figure 1.21 A lateral view of the right shoulder and upper extremity.

1. Trapezius m.
2. Long head of triceps brachii m.
3. Lateral head of triceps brachii m.
4. Lateral epicondyle of humerus
5. Olecranon of ulna
6. Acromion of scapula
7. Deltoid m.
8. Pectoralis major m.
9. Biceps brachii m.
10. Brachioradialis m.
11. Extensor carpi radialis longus and brevis mm.
12. Extensor digitorum m.

Figure 1.22 An anterior view of the right upper extremity.

1. Cephalic vein
2. Biceps brachii m.
3. Cubital fossa
4. Brachioradialis m.
5. Cephalic vein
6. Site for palpation of radial artery
7. Tendon of flexor carpi radialis m.
8. Tendon of palmaris longus m.
9. Thenar eminence
10. Metacarpophalangeal joint of thumb
11. Site of palpation of brachial artery
12. Basilic vein
15. Median antebrachial vein
16. Tendon of flexor digitorum superficialis m.
17. Styloid process of ulna
18. Hypothenar eminence

Figure 1.23 A posterior view of the right upper extremity.

1. Biceps brachii m.
2. Cubital fossa
3. Brachioradialis m.
4. Extensor carpi radialis longus m.
5. Extensor carpi ulnaris m.
6. Styloid process of radius
7. Tendon of extensor pollicus longus m.

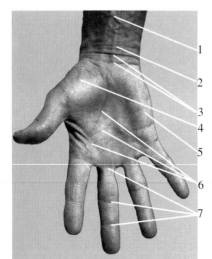

Figure 1.24 An anterior view of the right hand.

1. Tendon of flexor carpi radialis m.
2. Tendon of palmaris longus m.
3. Flexion creases on wrist
4. Thenar eminence
5. Hypothenar eminence
6. Flexion creases on palm of hand
7. Flexion creases on third digit

Figure 1.25 A posterior view of the right hand.

1. Styloid process of ulna
2. Position of extensor retinaculum
3. Tendons of extensor digitorum m.
4. Tendon of extensor digiti minimi m.
5. Metacarpophalangeal joint
6. Proximal interphalangeal joint
7. Distal interphalangeal joint
8. Nails

Figure 1.26 An anterior view of the right thigh.

1. Site of femoral triangle
2. Quadriceps femoris group of muscles
3. Adductor group of muscles
4. Rectus femoris m.
5. Vastus lateralis m.
6. Sartorius m.
7. Vastus medialis m.
8. Tendon of quadriceps femoris m.
9. Patella
10. Patellar ligament

Figure 1.27 A medial view of the right thigh.

1. Adductor magnus m.
2. Gracilis m.
3. Rectus femoris m.
4. Sartorius m.
5. Vastus medialis m.
6. Semimembranosus m.
7. Semitendinosus m.
8. Patella

Figure 1.28 A posterior view of the right thigh.

1. Gluteus maximus m.
2. Fold of buttock (gluteal fold)
3. Hamstring group of muscles
4. Vastus lateralis m.
5. Long head of biceps femoris m.
6. Semitendinosus m.
7. Gracilis m.
8. Popliteal fossa
9. Lateral head of gastrocnemius m.

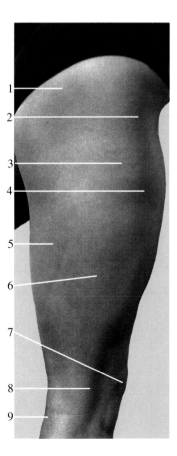

Figure 1.29 A lateral view of the right thigh.

1. Gluteus maximus m.
2. Tensor fasciae latae m.
3. Vastus lateralis m.
4. Rectus femoris m.
5. Biceps femoris m.
6. Iliotibial tract
7. Patella
8. Lateral epicondyle of femur
9. Popliteal fossa

Figure 1.30 An anterior view of the right leg and foot.

1. Patella
2. Patellar ligament
3. Tibialis anterior m.
4. Lateral malleolus of fibula
5. Medial malleolus of tibia
6. Site for palpation of dorsal pedis artery
7. Tendons of extensor digitorum longus m.
8. Tendon of extensor hallucis longus m.

Figure 1.31 A medial view of the right leg and foot.

1. Tibia
2. Medial head of gastrocnemius m.
3. Branch of great saphenous vein
4. Soleus m.
5. Tendo calcaneus
6. Medial malleolus of tibia
7. Calcaneus
8. Abductor hallucis m.
9. Longitudinal arch
10. Tendon of extensor hallucis longus m.
11. Head of first metatarsal bone

Figure 1.32 A posterior view of the right leg and foot.

1. Popliteal fossa
2. Lateral head of gastrocnemius m.
3. Medial head of gastrocnemius m.
4. Soleus m.
5. Peroneus longus m.
6. Tendo calcaneus
7. Peroneus brevis m.
8. Medial malleolus of tibia
9. Lateral melleolus of fibula
10. Calcaneus
11. Abductor digiti minimi m.
12. Plantar surface of foot

Figure 1.33 A lateral view of the right leg and foot.

1. Lateral head of gastrocnemius m.
2. Tibialis anterior m.
3. Peroneus longus m.
4. Soleus m.
5. Tendo calcaneus
6. Lateral malleolus of fibula
7. Calcaneus
8. Extensor digitorum brevis m.
9. Lateral surface of foot
10. Tendons of extensor digitorum longus m.

Cells

Cells are the basic structural and functional units of organization within the body. Although diverse, human cells have structural similarities including a *nucleus* containing a nucleolus, various *organelles* suspended in *cytoplasm*, and an enclosing *cell (plasma) membrane* (fig. 2.1).

The *nucleus* is the large spheroid body within a cell that contains the *chromatin*, *nucleolus*, and *nucleoplasm*—the genetic material of the cell. The nucleus is enclosed by a double membrane called the *nuclear membrane*, or *nuclear envelope*. The *nucleolus* is a dense, nonmembranous body composed of protein and RNA molecules. The chromatin consists of fibers of protein and DNA molecules. Prior to cellular division, the chromatin shortens and coils into rod-shaped *chromosomes*. Chromosomes consists of DNA and proteins called *histones*.

The *cytoplasm* of a cell is the medium of support between the nuclear membrane and the cell membrane. *Organelles* are minute membrane-bound structures within the cytoplasm of a cell that are concerned with specific functions. The cellular functions carried out by the organelles are referred to as *metabolism*. The principal organelles and their functions are listed in table 2.1. In order for cells to remain alive, metabolize, and maintain homeostasis, certain requirements must be met. These include having access to nutrients and oxygen, being able to eliminate wastes, and being maintained in a constant, protective environment.

The cell membrane is composed of phospholipid and protein molecules, which gives form to a cell and controls the passage of material into and out of a cell. More specifically, the proteins in the cell membrane provide: 1) structural support; 2) a mechanism of molecule transport across the membrane; 3) enzymatic control of chemical reactions; 4) receptors for hormones and other regulatory molecules; and 5) cellular markers (antigens), which identify the blood and tissue type. The carbohydrate molecules: 1) repel negative objects due to their negative charge; 2) act as receptors for hormones and other regulatory molecules; 3) form specific cell markers which enable like cells to attach and aggregate into tissues; and 4) enter into immune reactions.

The permeability of the cell membrane is a function of: 1) size of molecules; 2) solubility in lipids; 3) ionic charge of molecules; and 4) the presence of carrier molecules.

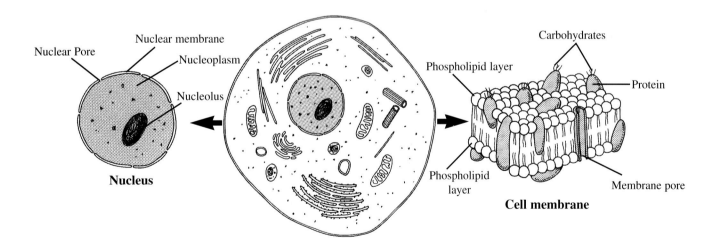

Figure 2.1 A cell and it's nucleus and cell (plasma) membrane.

Table 2.1 Structure and Function of Cellular Components

Component	Structure	Function
Cell (plasma) membrane	Composed of protein and phospholipid molecules	Provides form to cell and controls passage of materials into and out of cell
Cytoplasm	Fluid to jelly-like substance	Suspends organelles and provides a matrix in which chemical reactions occur
Endoplasmic reticulum	Interconnecting hollow channels	Provides supporting framework of cell; facilitates cell transport
Ribosomes	Granules of nucleic acid	Synthesize proteins
Mitochondria	Double-layered sacs with cristae	Production of ATP
Golgi apparatus	Flattened sacs with vacuoles	Synthesize carbohydrates and packages molecules for secretion
Lysosomes	Membrane-surrounded sacs of enzymes	Digest foreign molecules and worn cells
Centrosome	Mass of two rodlike centrioles	Organizes spindle fibers and assists mitosis
Vacuoles	Membranous sacs	Store and excrete substances within the cytoplasm
Fibrils and microfibrils	Protein strands	Support cytoplasm and transport materials
Cilia and flagella	Cytoplasmic extensions from cell	Movement of particles along cell surface or move cell
Nucleus	Nuclear membrane, nucleolus, and chromatin (DNA)	Directs cell activity; forms ribosomes

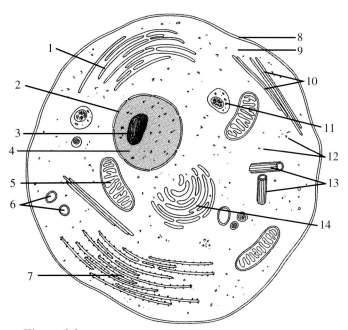

Figure 2.2 A typical animal cell.

1. Smooth endoplasmic reticulum
2. Nuclear membrane
3. Nucleolus
4. Nucleoplasm
5. Mitochondrion
6. Vesicles
7. Rough endoplasmic reticulum
8. Cell membrane
9. Cytoplasm
10. Microtubules
11. Lysosome
12. Ribosomes
13. Centrioles
14. Golgi apparatus

Nuclear pores

Photo courtesy of Scott C. Miller

Figure 2.3 An electron micrograph of a freeze fractured nuclear envelope showing the nuclear pores. (X1000)

Figure 2.4 An electron micrograph of various organelles.

1. Nucleus   3. Mitochondrion
2. Centrioles   4. Golgi apparatus

Figure 2.5 An electron micrograph of centrioles. The centrioles are positioned at right angles to one another.

1. Centriole   2. Centriole

Figure 2.6 An electron micrograph of cilia showing the characteristic "9 + 2" arrangement of microtubules in the cross sections.

1. Microtubules

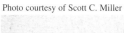

Figure 2.7 An electron micrograph showing the difference between a microvillus and a cilium.

1. Microvillus   2. Cilium

Figure 2.8 An electron micrograph of lysosomes.

1. Nucleus
2. Lysosomes

Figure 2.9 An electron micrograph of a mitochondrion.

1. Outer membrane
2. Crista
3. Inner membranes

Figure 2.10 An electron micrograph of smooth endoplasmic reticulum from the testis.

Figure 2.11 An electron micrograph of rough endoplasmic reticulum.

1. Ribosomes   2. Cisternae

Figure 2.12 Rough endoplasmic reticulum secreting collagenous filaments to the outside of the cell.

1. Nucleus
2. Rough endoplasmic
   reticulum
3. Collagenous
   filaments
4. Cell membrane

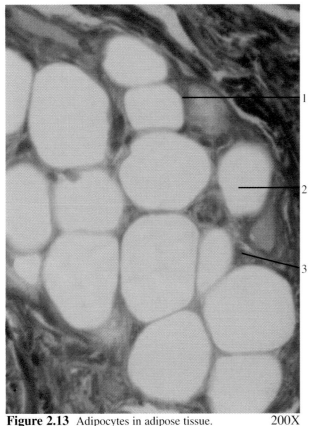

Figure 2.13 Adipocytes in adipose tissue.   200X

1. Cell membrane of adipocyte
2. Lipid-filled vacuole of adipocyte
3. Nucleus

Photo courtesy of Scott C. Miller

Figure 2.14 An electron micrograph of an osteocyte in cortical bone matrix.

1. Lacuna   3. Bone matrix
2. Osteocyte   4. Canaliculi

Photo courtesy of Scott C. Miller

Figure 2.15 Skeletal muscle cells (fibers).

1. Nucleus
2. Sarcolemma
3. Striations

Photo courtesy of Scott C. Miller

Figure 2.17 An electron micrograph of an erythrocyte (red blood cell).

Figure 2.16 An electron micrograph of a skeletal muscle myofibril, showing the striations.

1. Mitochondria   5. T-tubule
2. Z line   6. Sarcoplasmic reticulum
3. I band   7. H band
4. A band   8. Sacromere

Figure 2.18 Goblet cells within an intestinal gland (crypt of Lieberkühn) of small intestine. (X430)

1. Lumen of gland   2. Goblet cells

Figure 2.19 An electron micrograph of a capillary containing an erythrocyte.

1. Lumen of capillary   3. Endothelial cell
2. Erythrocyte   4. Nucleus of endothelial cell

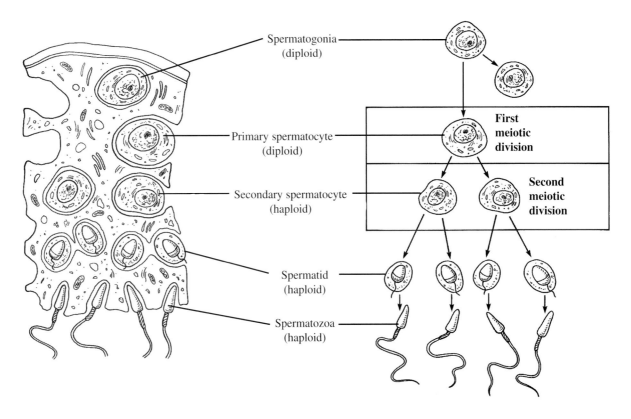

Figure 2.20 Spermatogenesis is the production of male gametes, or spermatozoa, through the process of meiosis.

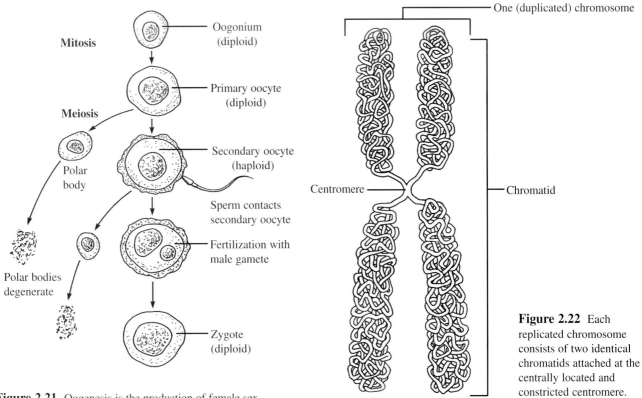

Figure 2.21 Oogenesis is the production of female sex gametes, or ova, through the process of meiosis.

Figure 2.22 Each replicated chromosome consists of two identical chromatids attached at the centrally located and constricted centromere.

Figure 2.23 Stages of mitosis.

Prophase

Each chromosome consists of two chromatids jointed by a centromere. Spindle fibers extend from each centriole.

Metaphase

The chromosomes are positioned at the equator. The spindle fibers from each centriole attach to the centromeres.

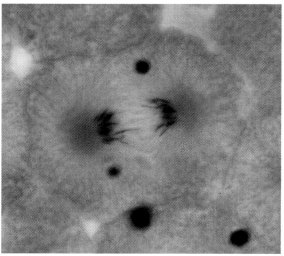

Anaphase

The centromeres split, and the sister chromatids separate as each is pulled to an opposite pole.

Telophase

The chromosomes lengthen and become less distinct. The cell membrane forms between the forming daughter cells.

(All photographs on this page are 800X.)

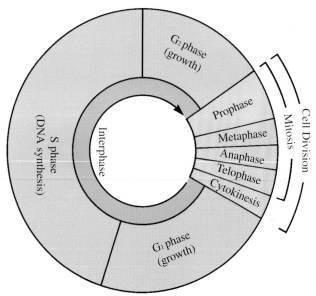

Figure 2.24 Animal cell cycle.

Histology

3

While it is true that cells comprise the basic structural and functional units of the body, the cells in a multicellular organism, such as a human, are so specialized that they do not function independently. *Tissues* are aggregations of similar cells that perform specific functions. *Histology* is the science concerned with the study of cells. Both *cytology*, the study of cells, and histology are actually microscopic anatomy. Although cytologists and histologists utilize many different techniques to study cells and tissues, basically only two kinds of microscopes are used the view the prepared specimens. *Light microscopy* is used for the general observation of cellular and tissue structure, and *electron microscopy* permits observation of the fine details of the specimens.

In electron microscopy, a beam of electrons is passed through an object in a procedure called *transmission electron microscopy (TEM)*, or the beam is reflected off the surface of an object in a procedure called *scanning electron microscopy (SEM)*. In both cases, the electron beam is magnified with electromagnets. The depth of focus of SEM is much greater than it is with SEM, producing a clear three-dimensional image of cellular or tissue structure. The magnification ability of SEM, however, is not as great as that of TEM.

The tissues of the body are classified into four principal types, determined by structure and function: 1) *epithelial tissue* covers body and organ surfaces, lines body and luminal (hollow portion of body tubes) cavities, and forms various glands; 2) *connective tissue* binds, supports, and protects body parts; 3) *muscle tissue* contracts to produce movements; and 4) *nervous tissue* initiates and transmits nerve impulses from one body part to another.

Epithelial tissue is classified by the number of layers of cells and the shape of the cells along the exposed surface. A *simple epithelial tissue* is made up of a single layer of cells. A *stratified epithelial tissue* is made up of layers of cells. The basic shapes of the exposed cells are: *squamous*, or flattened;

cuboidal, or cube-shaped; and *columnar*, or elongated.

Connective tissue is classified according to the characteristics of the *matrix*, or binding material between the similar cells. The classification of connective tissue is not exact, but the following is a commonly accepted scheme of classification:

A. Embryonic connective tissue

B. Connective tissue proper
1. Loose (areolar)
2. Dense regular
3. Dense irregular
4. Elastic
5. Reticular
6. Adipose

C. Cartilage
1. Hyaline
2. Fibrocartilage
3. Elastic

D. Bone

E. Vascular (blood) tissue

Muscle tissue is responsible for the movement of materials through the body, the movement of one part of the body with respect to another, and for locomotion. The three kinds of muscle tissue are *smooth, cardiac,* and *skeletal*. The fibers in all three kinds are adapted to contract in response to stimuli.

Nervous tissue is composed of *neurons*, which respond to stimuli and conduct impulses to and from all body organs, and *neuroglia*, which functionally support and physically bind neurons.

Figure 3.1 Simple squamous epithelium.   300X
1. Single layer of flattened cells

Figure 3.2 Simple cuboidal epithelium.   300X
1. Single layer of cells with round nuclei

Figure 3.3 Simple columnar epithelium.   300X

1. Single layer of cells with oval nuclei

Figure 3.4 Stratified squamous epithelium.   200X

1. Multiple layers of cells, which are flattened at the
   upper layer

Figure 3.5 Stratified cuboidal epithelium.   200X

1. Two layers of cells with round nuclei

Figure 3.6 Stratified columnar epithelium.   200X

1. Two layers of cells with oval nuclei

Figure 3.7 Stratified columnar epithelium.   200X

1. Cells are balloon-like at surface

Figure 3.8 Pseudostratified columnar   600X
epithelium.

1. Cilia
1. Pseudostratified columnar epithelium
2. Basement membrane

Figure 3.9 Adipose connective tissue. 200X

1. Adipocytes (adipose cells)

Figure 3.10 Loose connective tissue. 200X

Photo courtesy of Scott Miller

Figure 3.11 An electron micrograph of collagenous fibers in loose connective tissue.

1. Fibroblast
2. Cross section of bundle of collagenous fibers
3. Longitudinal section of bundle of collagenous fibers

Figure 3.12 Loose connective tissue stained for fibers. 200X

1. Elastic fibers (black)
2. Collagen fibers (pink)

Figure 3.13 Dense regular connective tissue. 200X

1. Nuclei of fibroblasts arranged in parallel rows

Figure 3.14 Electron micrograph of dense regular connective tissue. 200X

1. Collagenous fiber
2. Fibroblast

Figure 3.15 Dense irregular connective tissue.   200X

1. Epidermis
2. Dense irregular connective tissue (reticular layer of dermis)

Figure 3.16 Electron micrograph of   200X
dense irregular connective tissue.

1. Collagenous fibers

Figure 3.17 Dense irregular connective tissue.   300X

Figure 3.18 Reticular connective tissue.   200X

1. Reticular fibers

Figure 3.19 Hyaline cartilage.   200X

1. Chondrocytes
2. Hyaline cartilage

Figure 3.20 Elastic cartilage.   150X

1. Chondrocytes   3. Elastic fibers
2. Lacunae

Figure 3.21 Fibrocartilage.   150X

1. Chondrocytes arranged in a row

Figure 3.22 Cells of connective tissue.   600X

1. Eosinophils   3. Lymphocyte
2. Fibroblasts

Figure 3.23 Cells of connective tissue.   300X

1. Macrophages

Figure 3.24 Cells of connective tissue,   150X
special preparation.

1. Mast cells   2. Macrophages

Figure 3.25 Cells of connective tissue.   600X

1. Plasma cells

1. Zone of reserve cartilage
2. Zone of proliferation
3. Zone of hypertrophy
4. Zone of calcification
5. Zone of resorption
6. Calcified cartilage (blue)
7. Spicule of bone (red).

Figure 3.26 Endochondral   40X
bone formation.

Figure 3.27 Electron micrograph of   200X
bone tissue.

1. Interstitial lamellae   4. Lacunae
2. Lamellae   5. Osteon (haversian system)
3. Central canal
   (haversian canal)

Figure 3.28 SEM photomicrograph of   200X
bone tissue formation.

1. Bone mineral (calcium salts stain black)
2. Collagenous filament (distinct banding pattern)
3. Collagen secreting osteoblasts

Figure 3.29 Cross section of two osteons.   200X

1. Lacunae   3. Lamellae
2. Central (haversian) canals

Figure 3.30 Longitudinal section of   100X
osteons.

1. Central (haversian) canal
2. Perforating (Volkman's) canals
3. Central (haversian) canals

Figure 3.31 Spongy bone.   120X

1. Spicule of bone
2. Marrow (includes adipose cells)

Figure 3.32 Osteoblasts.　　　　　375X

1. Osteoblasts　　　　3. Osteoid
2. Bone　　　　　　　4. Osteocytes

Figure 3.33 Osteoclast.　　　　　375X

1. Osteocytes　　　　4. Osteoclast in Howship's
2. Bone　　　　　　　　 lacuna
3. Howship's lacuna

Figure 3.34 Longitudinal section of　　　250X
skeletal muscle tissue.

1. Skeletal muscle cells, note striations
2. Multiple nuclei in periphery of cell

Figure 3.35 Cross section of skeletal　　　400X
muscle tissue.

1. Skeletal muscle cells
2. Nuclei in periphery of cell
3. Endomysium (surrounds cells)
4. Perimysium (surrounds bundles of cells)

Figure 3.36 Attachment of skeletal　　　200X
muscle to tendon.

1. Skeletal muscle
2. Dense regular connective tissue (tendon).

Figure 3.37 Cardiac muscle tissue.　　　200X

1. Intercalated discs
2. Light-staining perinuclear sarcoplasm
3. Nucleus in center of cell

Figure 3.38 Smooth muscle tissue.   75X

1. Smooth muscle
2. Blood vessel

Figure 3.39 Partially teased smooth   250X
muscle tissue.

1. Nucleus of individual cell

Figure 3.40 Purkinje fibers.   400X

1. Purkinje fibers.

Figure 3.41 Neuron smear.   400X

1. Nuclei of surrounding   3. Nucleolus of neuron
   neuroglial cells   4. Dendrites of neuron
2. Nucleus of neuron

Figure 3.42 Cross section of a nerve.   40X

1. Perineurium   3. Endoneurium
2. Epineurium   4. Bundle of axons

Figure 3.43 Cross section of a nerve.   400X

1. Endoneurium   3. Perineurium
2. Axons   4. Epineurium

Figure 3.44 A nerve stained with osmium.　400X

1. Myelin sheath
2. Endoneurium
3. Axon

Figure 3.45 A longitudinal section of axons.　250X

1. Myelin sheath
2. Neurofibril nodes (nodes of Ranvier)

Figure 3.46 Neuromuscuscular　250X
junction.

1. Motor nerve
2. Motor end plates
3. Skeletal muscle fiber

Figure 3.47 Unmyelinated nerve.　100X

Figure 3.48 Cross section of the spinal cord.   7X

1. Posterior (dorsal) root of spinal nerve
2. Posterior (dorsal) horn (gray matter)
3. Spinal cord tract (white matter)
4. Anterior (ventral) horn (gray matter)

Figure 3.49 Motor neurons from spinal cord.   250X

1. Neuroglia cells
2. Dendrites
3. Nucleus

Figure 3.50 Pyramidal neurons from the   250X
cerebral cortex.

1. Apical dendrite
2. Other dendrites

Figure 3.51 Purkinje neurons from the   200X
cerebellum.

1. Molecular layer of cerebellar cortex
2. Granular layer of cerebellar cortex
3. Dendrites of Purkinje cell
4. Purkinje cell body

Integumentary System

The integumentary system consists of the *integument*, or *skin*, and its associated *hair*, *glands*, and *nails* (fig. 4.1). The skin is composed of an outer *epidermis* consisting of four or five layers, a *dermis* consisting of two layers, and a single-layered *hypodermis*, or *subcutaneous layer*.

The stratified squamous epithelium of the epidermis is divisible into five *strata*, or layers. From superficial to deep, they are the *stratum corneum*, the *stratum lucidum* (only in skin of the palms and soles), the *stratum granulosum*, the *stratum spinosum*, and the *stratum basale*. The strata basale and spinosum undergo mitosis (cell division) and are collectively called the *stratum germinativum*. Pigments are found in the stratum germinativum and the protein *keratin* is found in all but the deepest epidermal layers. Both are protective. The stratum corneum is *cornified* (hardened and scale-like) for further protection.

The dermis is divisible into the *stratum papillarosum* (papillary layer) and the *stratum reticularosum* (reticular layer). The hypodermis is the deep, binding layer.

The skin provides several important functions, including: 1) protection of the body from disease and external injury. Keratin and an acidic oily secretion on the surface protect the skin from water and microorganisms. Cornification protects against abrasion, and *melanin* (a dark pigment) is a barrier to UV light; 2) regulation of body fluids and temperatures by radiation, convection, and the antagonistic effects of sweating and shivering; 3) permits the absorption of some UV light, respiratory gases, steroids, and fat-soluble vitamins; 4) synthesizes melanin and keratin, which remain in the skin, and vitamin D, which is used elsewhere in the body; 5) sensory reception provided through cutaneous receptors throughout the dermis and hypodermis; and 6) development and growth of hair and certain exocrine glands.

Formed prenatally as invaginations of the epidermis into the dermis, hair, glands and nails provide protection to the skin. Each hair develops in a *hair follicle* and is protective against sunlight and mild abrasions. Integumentary glands are classified as *sebaceous* (oil secreting), *sudoriferous* (sweat), and *ceruminous* (wax-producing). (*Mammary glands* are specialized sweat glands that produce milk in a lactating female.) A nail protects the terminal end of each digit. The fingernails also aid in picking up objects and scratching.

Figure 4.1 The skin and certain epidermal structures.

1. Epidermis
2. Dermis
3. Hypodermis
4. Shaft of hair
5. Stratum corneum
6. Stratum basale
7. Sweat duct
8. Sensory receptor
9. Sweat duct
10. Sebaceous gland
11. Arrector pili muscle
12. Hair follicle
13. Apocrine sweat gland
14. Eccrine sweat gland
15. Bulb of hair
16. Adipose tissue
17. Cutaneous blood vessels

Figure 4.2 The gross structure of the skin and underlying fascia.

1. Epidermis   4. Fascia
2. Dermis   5. Muscle
3. Hypodermis

Figure 4.3 Skin.   10X

1. Lamellated (Pacinian) corpuscle   3. Dermis
2. Epidermis   4. Hypodermis

Figure 4.4 Epidermis and dermis.   75X

1. Stratum corneum   4. Stratum spinosum
2. Stratum lucidum   5. Stratum basale
3. Stratum granulosum   6. Dermis

Figure 4.5 Corpuscle of touch.   200X

1. Corpuscle of touch (Meissner's corpuscle)
2. Dermal papillae

Figure 4.6 Fingertip.   10X

1. Eponychium   4. Free border of nail
2. Nail plate   5. Hyponychium
3. Phalanges

Figure 4.7 Sweat gland.   75X

1. Excretory portion of sweat gland
2. Epidermis
3. Excretory duct of sweat gland
   (coiling toward surface)
4. Sebaceous gland
5. Secretory portion of sweat gland

Figure 4.8 Sweat gland.   200X

1. Secretory portion (large diameter with light–staining columnar cells)
2. Excretory portion (small diameter with dark–staining stratified cuboidal cells)

Figure 4.9 Sweat gland.   600X

1. Lumen of secretory portion   3. Lumen of excretory portion
2. Myoepithelial cell

40X

Figure 4.10 Hair follicle.

1. Epidermis
2. Sebaceous glands
3. Arrector pili muscle
4. Hair follicle
5. Hair follicle (oblique cut)

Figure 4.11 An electron micrograph of a hair emerging from a hair follicle.

1. Shaft of hair (note the scale–like pattern)
2. Hair follicle
3. Epithelial cell from stratum corneum

Skeletal System: Axial Portion   5

The skeletal system of an adult human is composed of approximately 206 bones—the number varies from person to person depending on genetic variations. Some adults have extra bones in the skull called *sutural (wormian) bones*. Additional bones may develop in tendons as the tendons move across a joint. Bones formed this way are called *sesamoid bones*, and the patella (kneecap) is an example.

The skeleton is divided into axial and appendicular portions. The *axial skeleton* consists of the bones that form the axis of the body and that support and protect the organs of the head, neck, and trunk. The axial skeleton includes the bones of the skull, auditory ossicles, hyoid bone, vertebral column, and rib cage.

The *appendicular skeleton* (see chapter 6) is composed of the bones of the upper and lower extremities and the bony girdles, which anchor the appendages to the axial skeleton. The appendicular skeleton includes the bones of the pectoral girdle, upper extremities, pelvic girdle, and lower extremities.

The mechanical functions of the bones of the skeleton include the support and protection of softer body tissues and organs. Also, certain bones function as levers during body movement. The metabolic functions of bones include *hemopoiesis*, or manufacture of blood cells, and mineral storage. Calcium and phosphorus are the two principal minerals stored within bone, and give bone its rigidity and strength.

The bones of the skeleton are classified into four principal types on the basis of shape rather than size. The four classes of bones are *long bones*, *short bones*, *flat bones* and *irregular bones* (fig. 5.1).

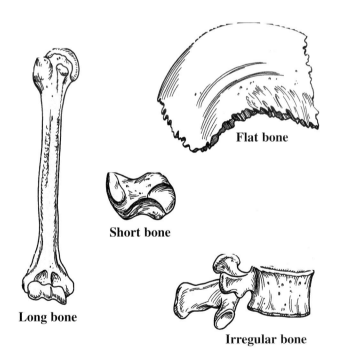

Figure 5.1 Shapes of bones.

Figure 5.2 A sagittal CT image of the head and neck.

Figure 5.3 An anterior view of the skeleton.

Figure 5.4 A posterior view of the skeleton.

1. Frontal bone	12. Pubis	23. Manubrium
2. Zygomatic bone	13. Carpal bones	24. Scapula
3. Mandible	14. Metacarpal bones	25. Costal cartilage
4. Cervical vertebra	15. Phalanges	26. Thoracic vertebra
5. Clavicle	16. Femur	27. Ulna
6. Body of sternum	17. Patella	28. Radius
7. Rib	18. Tarsal bones	29. Symphysis pubis
8. Humerus	19. Metatarsal bones	30. Fibula
9. Lumbar vertebra	20. Phalanges	31. Tibia
10. Ilium	21. Orbit	32. Calcaneus
11. Sacrum	22. Maxilla	

1. Parietal bone	10. Tibia	19. Ulna
2. Occipital bone	11. Fibula	20. Coccyx
3. Cervical vertebra	12.Tarsal bones	21. Carpal bones
4. Scapula	13. Mandible	22. Metacarpal bones
5. Humerus	14. Clavicle	23. Phalanges
6. Ilium	15. Thoracic vertebra	24. Metatarsal bones
7. Sacrum	16. Rib	25. Phalanges
8. Ischium	17. Lumbar vertebra	
9. Femur	18. Radius	

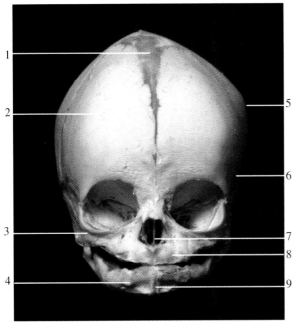

Figure 5.5 An anterior view of the fetal skull.

1. Anterior fontanel
2. Frontal bone
3. Zygomatic bone
4. Mandible
5. Parietal bone
6. Anterolateral fontanel
7. Nasal septum
8. Maxilla
9. Mental symphysis

Figure 5.6 A lateral view of the fetal skull.

1. Parietal bone
2. Area of lambdoidal suture
3. Occipital bone
4. Posterolateral fontanel
5. Temporal bone
6. Anterior fontanel
7. Frontal bone
8. Area of coronal suture
9. Anterolateral fontanel
10. Nasal bone
11. Sphenoid bone
12. Zygomatic bone
13. Maxilla
14. Mandible
15. Condylar process
16. External auditory meatus

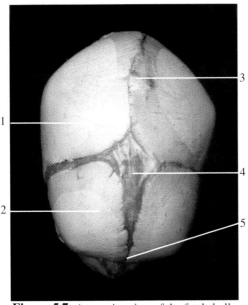

Figure 5.7 A superior view of the fetal skull.

1. Parietal bone
2. Frontal bone
3. Area of sagittal suture
4. Anterior fontanel
5. Synostosis

Figure 5.8 A posterior view of the fetal skull.

1. Parietal bone
2. Posterior fontanel
3. Occipital bone
4. Area of sagittal suture
5. Area of lambdoidal suture

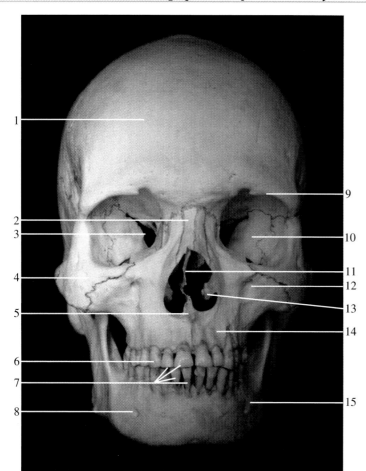

Figure 5.9 An anterior view of the skull.

1. Frontal bone
2. Nasal bone
3. Superior orbital fissure
4. Zygomatic bone
5. Vomer
6. Canine
7. Incisors
8. Mental foramen
9. Supraorbital margin
10. Sphenoid bone
11. Perpendicular plate of ethmoid bone
12. Infraorbital foramen
13. Inferior nasal concha
14. Maxilla
15. Mandible

Figure 5.10 A lateral view of the skull.

1. Coronal suture
2. Frontal bone
3. Lacrimal bone
4. Nasal bone
5. Zygomatic bone
6. Maxilla
7. Premolars
8. Molars
9. Mandible
10. Parietal bone
11. Squamosal suture
12. Temporal bone
13. Lambdoidal suture
14. Occipital bone
15. External acoustic meatus
16. Condylar process of mandible
17. Mastoid process of temporal bone
18. Mandibular notch
19. Coronoid process of mandible
20. Angle of mandible

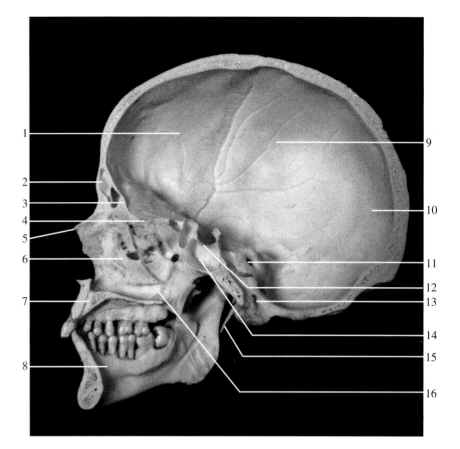

Figure 5.11 A sagittal view of the skull.

1. Frontal bone
2. Frontal sinus
3. Crista galli of ethmoid bone
4. Cribriform plate of ethmoid bone
5. Nasal bone
6. Nasal concha
7. Maxilla
8. Mandible
9. Parietal bone
10. Occipital bone
11. Internal acoustic meatus
12. Sella turcica
13. Hypoglossal canal
14. Sphenoidal sinus
15. Styloid process of temporal bone
16. Vomer

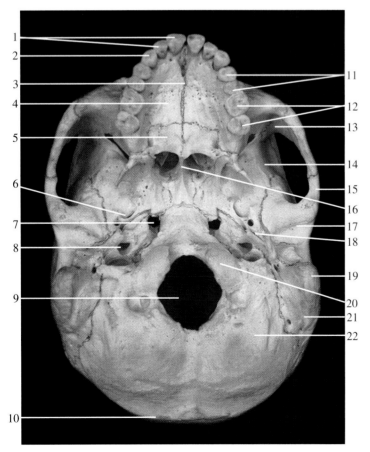

Figure 5.12 An inferior view of the skull.

1. Incisors
2. Canine
3. Intermaxillary suture
4. Maxilla
5. Palatine bone
6. Foramen ovale
7. Foramen lacerum
8. Carotid canal
9. Foramen magnum
10. Superior nuchal line
11. Premolars
12. Molars
13. Zygomatic bone
14. Sphenoid bone
15. Zygomatic arch
16. Vomer
17. Mandibular fossa
18. Styloid process of temporal bone
19. Mastoid process of temporal bone
20. Occipital condyle
21. Temporal bone
22. Occipital bone

Nasal bone
Zygomatic bone
Maxilla

Figure 5.13 Bones of the left facial region.

1. Orbital process of zygomatic bone	7. Lacrimal bone
2. Zygomatic bone	8. Frontal process of maxilla
3. Infraorbital foramen	9. Nasal bone
4. Maxilla	10. Inferior nasal concha
5. Molar	11. Canine
6. Premolars	12. Incisors

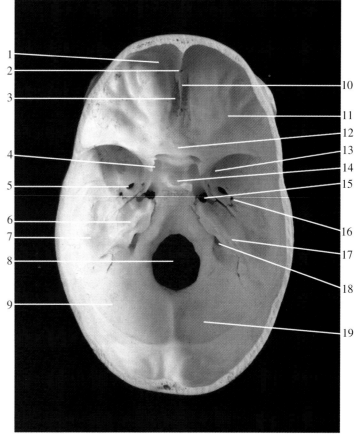

Figure 5.14 A superior view of the cranium.

1. Frontal bone	10. Crista galli of ethmoid bone
2. Foramen cecum	11. Anterior cranial fossa
3. Cribriform plate of ethmoid bone	12. Sphenoid bone
4. Optic canal	13. Foramen rotundum
5. Foramen ovale	14. Sella turcica of sphenoid bone
6. Petrous portion of temporal bone	15. Foramen lacerum
7. Temporal bone	16. Foramen spinosum
8. Foramen magnum	17. Internal acoustic meatus
9. Occipital bone	18. Jugular foramen
	19. Posterior cranial fossa

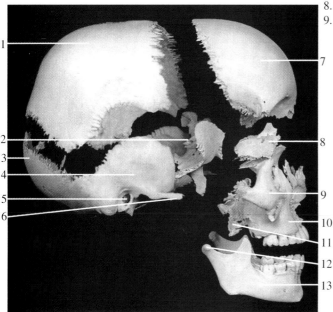

Figure 5.15 A lateral view of a disarticulated skull.

1. Parietal bone
2. Sphenoid bone
3. Occipital bone
4. Temporal bone
5. External acoustic meatus
6. Zygomatic process of temporal bone
7. Frontal bone
8. Ethmoid bone
9. Zygomatic bone
10. Maxilla
11. Palatine bone
12. Condylar process of mandible
13. Mandible

Figure 5.16 The orbit of the skull.

1. Ethmoid bone
2. Nasal bone
3. Lacrimal bone
4. Inferior nasal concha
5. Vomer
6. Maxilla
7. Frontal bone

8. Optic foramen
9. Superior orbital fissure
10. Sphenoid bone
11. Temporal bone
12. Inferior orbital fissure
13. Lacrimal foramen
14. Zygomatic bone

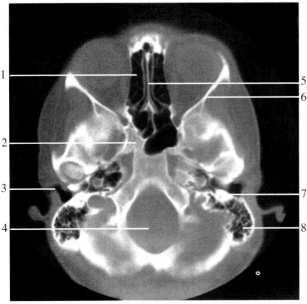

Figure 5.17 A CT transaxial section through the skull.

1. Nasal cavity
2. Sphenoid bone
3. External acoustic meatus
4. Foramen magnum

5. Nasal septum
6. Wall of bony orbit
7. Middle-ear chamber
8. Mastoidal sinus

Figure 5.18 A CT image of the nasal cavity and paranasal sinuses.

1. Eyeball in orbit
2. Nasal cavity
3. Crista galli
4. Frontal sinus
5. Perpendicular plate of ethmoid bone

6. Ethmoidal sinus
7. Maxillary sinus
8. Inferior nasal concha

Figure 5.19 An anterior view of the sphenoid bone.

1. Greater wing of sphenoid bone
2. Optic foramen
3. Opening into sphenoidal sinus
4. Medial pterygoid plate
5. Lesser wing of sphenoid bone
6. Superior orbital fissure
7. Orbital surface of greater wing of sphenoid bone
8. Foramen rotundum
9. Pterygoid canal
10. Lateral pterygoid plate

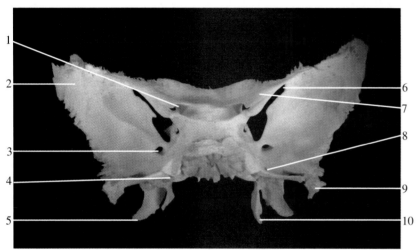

Figure 5.20 A posterior view of the sphenoid bone.

1. Optic foramen
2. Greater wing of sphenoid bone
3. Foramen rotundum
4. Pterygoid canal
5. Lateral pterygoid plate
6. Superior orbital fissure
7. Lesser wing of sphenoid bone
8. Foramen ovale
9. Spine of sphenoid bone
10. Medial pterygoid plate

Figure 5.21 An inferior view of the occipital bone.

1. External occipital protuberance
2. Foramen magnum
3. Occipital condyle
4. Pharyngeal tubercle

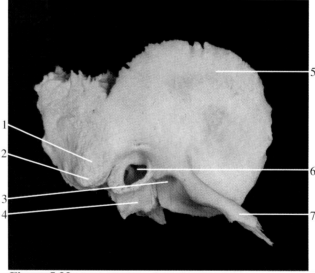

Figure 5.22 A lateral view of the temporal bone.

1. Mastoid portion of temporal bone
2. Mastoid process
3. Mandibular fossa
4. Tympanic portion of temporal bone
5. Squamous portion of temporal bone
6. External acoustic meatus
7. Zygomatic process of temporal bone

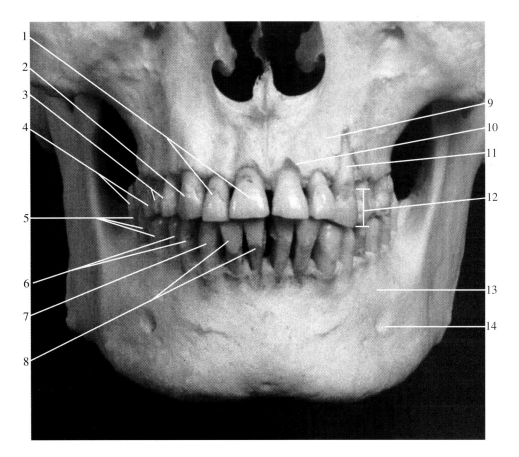

Figure 5.23 An anterior view of the jaws and teeth.

1. Superior incisors
2. Superior canine
3. Superior premolars
4. Superior molars
5. Inferior molars
6. Inferior premolars
7. Inferior canine
8. Inferior incisors
9. Maxilla
10. Dental alveolus
11. Root of tooth
12. Crown of tooth
13. Mandible
14. Mental foramen

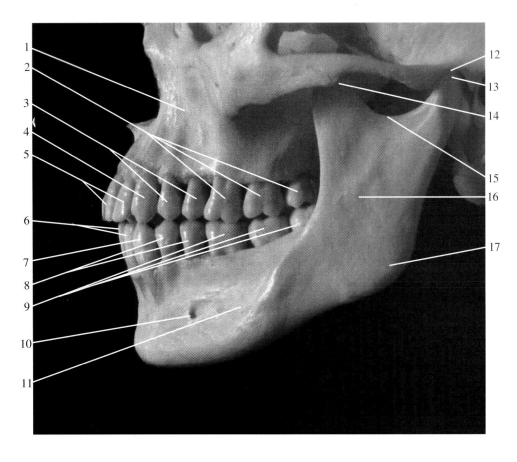

Figure 5.24 A lateral view of the jaws and teeth.

1. Maxilla
2. Superior molars
3. Superior premolars
4. Superior canine
5. Superior incisors
6. Inferior incisors
7. Inferior canine
8. Inferior premolars
9. Inferior molars
10. Mental foramen
11. Body of mandible
12. Temporomandibular joint
13. Condylar process of mandible
14. Coronoid process of mandible
15. Mandibular notch
16. Ramus of mandible
17. Angle of mandible

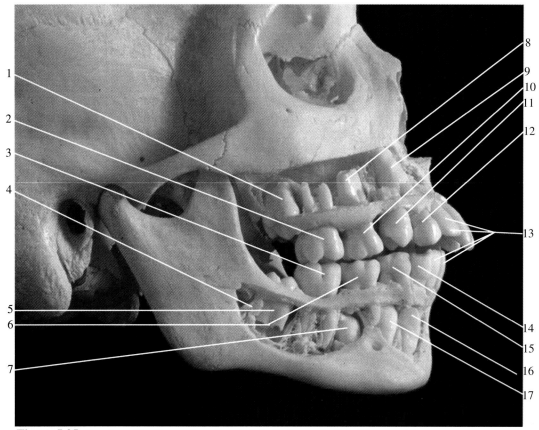

Figure 5.25 Eruption of teeth seen in a dissected skull of a youth (9 to 12 years old).

1. Permanent second molar
2. Permanent first molar
3. Permanent first molar
4. Permanent third molar
5. Permanent second molar
6. Deciduous second premolar

7. Permanent second premolar
8. Permanent second premolar
9. Permanent canine
10. Deciduous second premolar
11. Permanent first premolar
12. Deciduous canine

13. Incisors
14. Deciduous canine
15. Deciduous first premolar
16. Permanent canine
17. Permanent first premolar

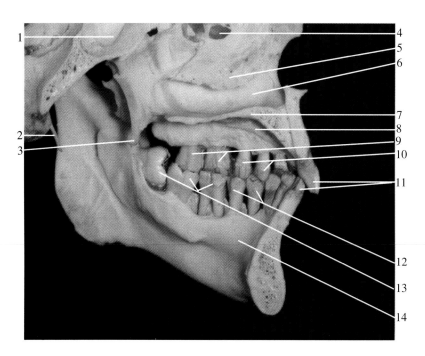

Figure 5.26 A medial view
of the jaws and teeth.

1. Sphenoidal sinus
2. Styloid process of temporal bone
3. Medial pterygoid process
4. Ethmoidal sinus
5. Inferior nasal concha
6. Inferior meatus
7. Hard palate
8. Maxilla
9. Superior molars
10 Superior premolars
11. Incisors
12. Inferior premolars
13. Inferior molars
    (note impacted wisdom tooth)
14. Mandible

Figure 5.27 The curvature of the vertebral column.

1. Cervical curvature
2. Thoracic curvature
3. Lumbar curvature
4. Pelvic curvature

Figure 5.28 A posterior view of the vertebral column.

1. Atlas
2. Axis
3. Seventh cervical vertebra
4. First thoracic vertebra
5. Twelfth thoracic vertebra
6. First lumbar vertebra
7. Fifth lumbar vertebra
8. Sacroiliac joint
9. Cervical vertebrae
10. Thoracic vertebrae
11. Lumbar vertebrae
12. Sacrum
13. Coccyx

Figure 5.29 Representative vertebrae.

1. Spinous process
2. Lamina
3. Spinous process
   (note that it is bifid)
4. Predicle
5. Vertebral foramen
6. Transverse foramen
7. Bodies of vertebrae
8. Superior articular surface
9. Transverse process
10. Costal fovea of transverse process
11. Inferior costal fovea

Figure 5.30 Superior views of (a) the atlas and (b) the axis.

1. Vertebral canal
2. Transverse foramen
3. Spinous process of atlas
4. Lamina of neural arch
5. Transverse process of atlas
6. Anterior arch of atlas
7. Spinous process of axis
8. Vertebral canal
9. Superior articular facet for atlas
10. Dens (odontoid process)

Figure 5.31 A radiograph of the lumbar vertebrae.

1. T12
2. Body of L1
3. Spinous process of L2
4. Intervertebral disc
5. Transverse process of L4
6. Lamina of L5
7. Sacrum
8. Sacroiliac joint

Figure 5.32 A radiograph of the cervical vertebrae.

1. Occipital condyle
2. Atlas
3. Axis
4. Intervetebral disc
5. Body of C3
6. Intervetebral foramen
7. Spinous process of C5
8. Spinous process of C6
9. Body of C7

Figure 5.33 A radiograph of the lumbar region showing the locations of pellets from a shotgun wound.

Figure 5.34 An anterior view of the sacrum and coccyx.

1. Sacrum
2. Coccyx
3. Base of sacrum
4. Ala
5. Transverse line
6. Anterior sacral foramina
7. Apex of sacrum
8. Coccygeal cornu

Figure 5.35 A posterior view of the sacrum and coccyx.
1. Superior portion of sacral canal
2. Ala
3. Posterior sacral foramina
4. Coccygeal cornu
5. Superior articular process
6. Auricular surface
7. Median sacral crest
8. Coccyx

Figure 5.36 An anterior view of the rib cage.

1. True ribs (seven pairs)
2. False ribs (five ribs)
3. Floating ribs (inferior two pairs of false ribs)
4. Jugular notch
5. Manubrium
6. Body of sternum
7. Xiphoid process
8. Costal cartilage
9. Twelfth thoracic vertebra
10. Twelfth rib

Figure 5.37 A posterior view of the rib cage.

1. Intercostal spaces
2. Transverse process of thoracic vertebra
3. Head of rib
4. Angle of rib
5. Body of rib

Figure 5.38 A rib.

1. Superior border
2. Body of rib
3. Articulation site for costal cartilage
4. Internal surface
5. Costal groove
6. Angle of rib
7. Neck of rib
8. Articular surface
9. Head of rib

Skeletal System: Appendicular Portion   6

The structure of the *pectoral girdle* and *upper extremities* of the appendicular skeleton is adaptive for freedom of movement and extensive muscle attachment. The structure of the *pelvic girdle* and *lower extremities* is adaptive for support and locomotion.

The pectoral girdle is composed of two *scapulae* and two *clavicles*. The clavicles attach the pectoral girdle to the axial skeleton at the *sternum*. The bones of each upper extremity are the *humerus*, *ulna*, *radius*, eight *carpal bones*, five *metacarpal bones*, and fourteen *phalanges*. In addition, two or more *sesamoid bones* at the interphalangeal joints are usually present.

The pelvic girdle, or *pelvis*, is formed by two *os coxae* (hip bones) which are united anteriorly by the *symphysis pubis*. The pelvic girdle is attached posteriorly to the sacrum of the vertebral column. Each *os coxa* consists of three separate bones: the *ilium*, the *ischium*, and the *pubis*. These bones are fused in an adult. The bones of each lower extremity are the *femur*, *patella*, *tibia*, *fibula*, seven *tarsal bones*, five *metatarsal bones*, and fourteen *phalanges*. In addition, two or more sesamoid bones at the interphalangeal joints are usually present.

The various surface features used to identify specific bones are presented in table 6.1.

Table 6.1 Surface Features Used to Identify Specific Bones

Articulating surfaces	Description
Condyle	A large, rounded, articulating knob
Facet	A flattened or shallow articulating surface
Head	A prominent, rounded, articulating end of a bone
Nonarticulating prominences	
Crest	A narrow, ridgelike projection
Epicondyle	A projection above a condyle
Process	Any marked bony prominence
Spine	A sharp, slender process
Trochanter	A massive process found only on the femur
Tubercle	A small rounded process
Tuberosity	A large roughened process
Depressions and openings	
Alveolus	A deep pit or socket
Fissure	A narrow, slitlike opening
Foramen (plural, *foramina*)	A rounded opening through a bone
Fossa	A flattened or shallow surface
Fovea	A small pit or depression; some are articular and some are not
Meatus, or canal	A tubelike passageway through a bone
Sinus	A cavity or hollow space in a bone
Sulcus	A groove in a bone

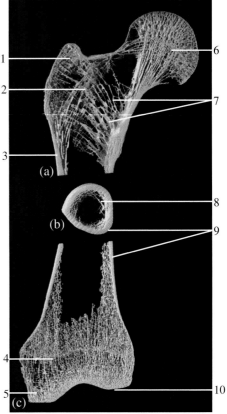

Figure 6.1 Sections of a human femur showing the outer compact bone and inner spongy bone. Note the trajectorial (stress) lines of the trabeculae of spongy bone. (a) A coronal section of the proximal end of the femur, (b) a cross section of the body of the femur, and (c) a coronal section of the distal end of the femur.

1. Greater trochanter
2. Spongy bone
3. Compact bone
4. Remnant of epiphyseal plate
5. Femoral epicondyle
6. Head of femur
7. Trabeculae of spongy bone
8. Spongy bone surrounding medullary cavity
9. Compact bone
10. Condyle of femur

Figure 6.3 An anterior view of the left scapula.

1. Superior angle
2. Medial (vertebral) border
3. Inferior angle
4. Acromion
5. Coracoid process
6. Glenoid fossa
7. Infraglenoid tubercle
8. Subscapular fossa
9. Lateral (axillary) border

Figure 6.2 Sections of a human tibia showing the outer compact bone and the inner spongy bone. Note the trajectorial stress lines of the trabeculae of spongy bone. (a) A coronal section of the proximal end of the tibia, (b) a cross section of the body of tibia, and (c) a coronal section of the distal end of the tibia.

1. Intercondylar eminences
2. Reminant of epiphyseal plate
3. Spongy bone
4. Compact bone
5. Medial malleolus
6. Lateral condyle
7. Trabeculae of spongy bone
8. Spongy bone

Figure 6.4 A posterior view of the left scapula.

1. Acromion
2. Glenoid fossa
3. Lateral (axillary) border
4. Superior angle
5. Supraspinous fossa
6. Spine
7. Infraspinous fossa
8. Medial (vertebral) border
9. Inferior angle

Figure 6.5 An inferior view of the clavicle.

1. Conoid tubercle
2. Acromial extremity
3. Sternal extremity

Figure 6.6 The right humerus. (a) An anterior view and (b) a posterior view.

1. Greater tubercle
2. Intertubercular groove
3. Lesser tubercle
4. Deltoid tuberosity
5. Anterior surface of humerus
6. Lateral supracondylar ridge
7. Lateral epicondyle
8. Capitulum
9. Head of humerus
10. Surgical neck
11. Posterior surface of humerus
12. Olecranon fossa
13. Coronoid fossa
14. Medial epicondyle
15. Trochlea
16. Greater tubercle
17. Anatomical neck
18. Lateral epicondyle

Figure 6.7 An anterior view of the left ulna and radius.

1. Olecranon
2. Interosseous margin
3. Head of ulna
4. Styloid process of ulna
5. Trochlear notch
6. Head of radius
7. Neck of radius
8. Radial tuberosity
9. Interosseous margin
10. Ulnar notch of radius
11. Styloid process of radius

Figure 6.8 A posterior view of the left ulna and radius.

1. Head of radius
2. Neck of radius
3. Interosseous margin
4. Styloid process of radius
5. Olecranon
6. Radial notch of ulna
7. Interosseous margin
8. Head of ulna
9. Styloid process of ulna

Figure 6.9 A radiograph of the left elbow region, posterior view.

1. Medial epicondyle of humerus
2. Olecranon
3. Trochlea
4. Radial notch of ulna
5. Tuberosity of ulna
6. Ulna
7. Humerus
8. Olecranon fossa of humerus
9. Lateral epicondyle of humerus
10. Capitulum
11. Articular surface of radius
12. Head of radius
13. Neck of radius
14. Tuberosity of radius
15. Radius

Figure 6.10 A radiograph showing fractures of the ulna and radius of a ten-year-old child. Notice the distal epiphyseal plates of the ulna and the radius.

1. Ulna
2. Radius

Figure 6.11 A radiograph showing a fracture of the distal portion of the radius and medial displacement of the antebrachium.

1. Ulna
2. Radius
3. Site of fracture
4. Styloid process of radius
5. Carpal bones
6. First metacarpal bone

Figure 6.12 A posterior view of the left wrist and hand.

1. Phalanges of fifth digit
2. Metacarpal bones (first-fifth)
3. Carpal bones
4. Distal phalanx of third digit
5. Middle phalanx of third digit
6. Proximal phalanx of third digit
7. Head of phalanx
8. Body of phalanx
9. Base of phalanx
10. Hamate bone
11. Capitate bone
12. Trapezoid bone
13. Trapezium bone
14. Triquetrum bone
15. Lunate bone
16. Scaphoid bone

Figure 6.13 A radiograph of the right wrist and hand, anteroposterior view.

1. Phalanges of fifth digit
2. Fifth metacarpal bone
3. Hamate bone
4. Capitate bone
5. Triquetrum and pisiform bones
6. Styloid process of ulna
7. Distal radioulnar joint
8. Ulna
9. Distal phalanx of third digit
10. Middle phalanx of third digit
11. Proximal phalanx of third digit
12. Distal phalanx of pollex (thumb)
13. Proximal phalanx of pollex (thumb)
14. Sesamoid bone
15. First metacarpal bone
16. Trapezoid bone
17. Trapezium bone
18. Scaphoid bone
19. Capitate bone
20. Lunate bone
21. Styloid process of radius
22. Radius

Figure 6.14 A photograph (a) and a radiograph (b) showing polydactyly, having extra digits. Polydactyly is a common congenital deformity of the hand, although it also occurs in the foot. Notice that the carpal bones are still cartilaginous in a newborn and do not show up on a radiograph.

Figure 6.15 An anterior view of the articulated pelvic girdle showing the two coxal bones, the sacrum, and the two femora.

1. Lumbar vertebra
2. Intervertebral disc
3. Ilium
4. Iliac fossa
5. Anterior superior iliac spine
6. Head of femur
7. Greater trochanter
8. Symphysis pubis
9. Crest of the ilium
10. Sacroiliac joint
11. Sacrum
12. Pelvic brim
13. Acetabulum
14. Pubic crest
15. Obturator foramen
16. Ischium
17. Pubic angle

Figure 6.16 A posterior view of the articulated pelvic girdle showing the two coxal bones, the sacrum, and the two femora.

1. Lumbar vertebra
2. Crest of ilium
3. Ilium
4. Sacrum
5. Greater sciatic notch
6. Coccyx
7. Head of femur
8. Greater trochanter
9. Intertrochanteric crest
10. Lesser trochanter
11. Sacroiliac joint
12. Acetabulum
13. Obturator foramen
14. Ischium
15. Pubis

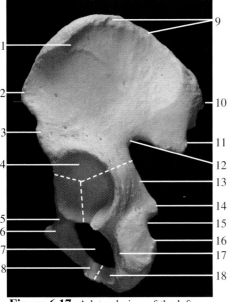

Figure 6.17 A lateral view of the left os coxa.

1. Ilium
2. Anterior superior iliac spine
3. Anterior inferior iliac spine
4. Acetabulum
5. Superior ramus of pubis
6. Pubis
7. Obturator foramen
8. Inferior ramus of pubis
9. Crest of the ilium
10. Posterior superior iliac spine
11. Posterior inferior iliac spine
12. Greater sciatic notch
13. Ischial body
14. Ischial spine
15. Lesser sciatic notch
16. Ischial tuberosity
17. Ischium
18. Ischial ramus

Figure 6.18 A medial view of the left os coxa.

1. Posterior superior iliac spine
2. Auricular surface
3. Posterior inferior iliac spine
4. Greater sciatic notch
5. Ischial spine
6. Ischium
7. Obturator foramen
8. Ischial ramus
9. Ilium
10. Iliac fossa
11. Anterior superior iliac spine
12. Anterior inferior iliac spine
13. Superior ramus of pubis
14. Pubis
15. Pubic crest
16. Pubic ramus

Figure 6.19 A radiograph of the left os coxa and femur, anterior view.

1. Ilium
2. Pelvic brim
3. Head of femur
4. Acetabulum
5. Neck of femur

6. Greater trochanter
7. Obturator foramen
8. Ischium
9. Femur

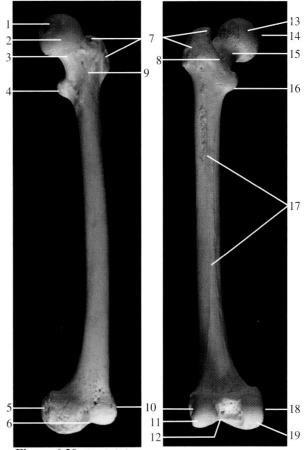

Figure 6.20 The left femur.

1. Fovea capitis femoris
2. Head
3. Neck
4. Lesser trochanter
5. Medial epicondyle
6. Patellar surface
7. Greater trochanter
8. Intertrochanteric crest
9. Intertrochanteric line
10. Lateral epicondyle

11. Lateral condyle
12. Intercondylar fossa
13. Head
14. Fovea capitis femoris
15. Neck
16. Lesser trochanter
17. Linea aspera on shaft (body) of femur
18. Medial epicondyle
19. Medial condyle

Figure 6.21 Radiograph of the left knee region, lateral view.

1. Femur
2. Intercondylar eminences of tibia
3. Tibia
4. Fibula
5. Patella
6. Lateral epicondyle of femur
7. Condyle of femur
8. Lateral epicondyle of tibia
9. Tibial tuberosity

Figure 6.22 An anterior view of the left patella, tibia, and fibula.

1. Base of patella
2. Apex of patella
3. Medial condyle
4. Tibial tuberosity
5. Anterior crest of tibia
6. Body (shaft) of tibia
7. Medial malleolus
8. Intercondylar tubercles
9. Lateral condyle
10. Tibial articular facet of fibula
11. Head of fibula
12. Neck of fibula
13. Body (shaft) of fibula
14. Lateral malleolus

Figure 6.23 A posterior view of the left tibia and fibula.

1. Intercondylar tubercles
2. Lateral condyle of tibia
3. Lateral epicondyle of tibia
4. Head of fibula
5. Neck of fibula
6. Body (shaft) of fibula
7. Lateral malleolus
8. Medial condyle of tibia
9. Medial epicondyle of tibia
10. Fibular articular facet of tibia
11. Body (shaft) of tibia
12. Fibular notch of tibia
13. Medial malleolus

Figure 6.24 A severe fracture of the leg and ankle (talus bone). In this patient, the trauma was so extensive amputation of the leg was necessary.

Figure 6.25 Radiograph showing fractures of the tibia and fibula.

1. Tibia
2. Fibula

Figure 6.26 An inferior view of the left foot.

1. Phalanges of fifth digit
2. Metatarsal bones (first–fifth)
3. Tarsal bones
4. Distal phalanx of first digit
5. Proximal phalanx of first digit
6. Head of first metatarsal bone
7. Body of first metatarsal bone
8. Base of first metatarsal bone
9. Medial cuneiform
10. Intermediate cuneiform
11. Lateral cuneiform
12. Navicular bone
13. Cuboid bone
14. Talus bone
15. Calcaneus bone
16. Tuberosity of calcaneus

Figure 6.27 An MR image of the ankle region, medial view.

1. Navicular bone
2. Medial cuneiform
3. First metatarsal bone
4. Tibia
5. Tendo calcaneus
6. Tibiotalar joint
7. Talus
8. Calcaneus

Figure 6.28 A radiograph of the left foot, medial view.

1. Tibia
2. Fibula (superimposed)
3. Tibiotalar joint
4. Talus
5. Calcaneus
6. Navicular bone
7. First cuneiform
8. Metatarsal bones
9. Phalanges
10. Sesamoid bone

Articulations

The structure of a joint determines its range of movement. Not all joints are flexible, however, and as one part of the body moves, other joints remain rigid to stabilize the body and maintain balance. *Arthrology* is the study of joints and *kinesiology* is the study of body movement, or the functional relationship between the skeleton, joints, muscles, and innervation (nerve supply) as they work together to produce coordinated movement.

The joints of the body are structurally classified into three basic kinds:

1. **Fibrous** – fibrous connective tisues join the skeletal structures; the joints lack a joint capsule and no movement is possible.

2. **Cartilaginous** – fibrocartilage or hyaline cartilage joins the skeletal structures; the joints lack a joint capsule, but in some, slight movement is possible.

3. **Synovial** – joint capsules containing synovial fluid are present between the articulating bones; articular cartilages and ligaments supporting the articulating bones are also present, which permit freedom of movement. Synovial joints are the freely moveable joints of the body.

Table 7.1 Movements Permitted at Synovial Joints

Type of movement	Description
Angular movement	Increase or decrease the joint angle
Flexion	Decreasing the angle between two bones
Extension	Increasing the angle between two bones
Hyperextension	Excessive extension beyond 180° (angle of anatomical position)
Dorsiflexion	Bending the foot toward the tibia
Plantar flexion	Bending the foot away from the tibia
Abduction	Movement of a body part away from the axis of the body, or away from the midsagittal plane, in a lateral direction
Inversion	Movement of the sole of the foot inward, or medially
Eversion	Movement of the sole of the foot outward, or laterally
Circular movement	
Rotation	Turning of a bone on its own axis
Circumduction	Movement of a body segment in a circular, conelike motion

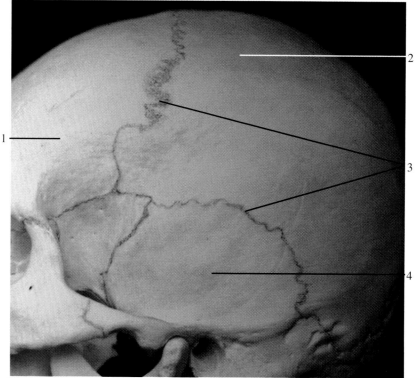

Figure 7.1 A suture located between cranial bones is a type of fibrous joint.

1. Frontal bone
2. Parietal bone
3. Sutures
4. Temporal bone

54

Table 7.2 Types of Joints

Type	Structure and movement	Example
Fibrous	Fibrous connective tissue joins the skeletal structures	
Suture	Frequently serrated edges of articulating bones, separated by a thin layer of fibrous tissue; no movement	Sutures between bones, cranial bones
Syndesmoses	Articulating bones bound by an interosseous ligament; slight movement	Joints between tibia-fibula and radius-ulna
Gomphoses	Periodontal ligament binding teeth into aveoli of bone; no movement	Teeth secured in the dental alveoli (teeth sockets)
Cartilaginous	Fibrocartilage or hyaline cartilage joins the skeletal structures	
Symphyses	Thin pad of fibrocartilage between articulating bones; slight movement	Symphysis pubis, sacroiliac joint and intervertebral joints
Synchondroses	Mitotically active hyaline cartilage between skeletal structures; no movement	Epiphyseal plates between diaphysis and epiphyses of long bones
Synovial	Joint capsule between articulating bones, containing synovial fluid; extensive movement	
Gliding	Flattened or slightly curved articulating surfaces; sliding movement	Intercarpal and intertarsal joints
Hinge	Concave surface of one bone articulates with a depression of another; bending motion in one plane	Humeroulnar (elbow) and tibiofemoral (knee) joints; interphalangeal (finger) joints of digits
Pivot	Conical surface of one bone articulates with a depression of another; rotation about a central axis; rotational movement	Atlantoaxial joint; proximal radioulnar joint
Condyloid	Oval condyle of one bone articulates with elliptical cavity of another; biaxial movement	Radiocarpal (wrist) joint
Saddle	Concave and convex surface on each articulating bone; wide range of movement	Carpometacarpal joint at the base of the thumb
Ball-and-socket	Rounded convex surface of one bone articulates with cuplike socket of another; movement in all planes and rotation	Humeroscapular (shoulder) and coxal (hip) joints

Figure 7.2 A symphysis is a type of cartilaginous joint. The symphysis pubis can be readily seen in this radiograph of the pelvic girdle. Note the head of the femur articulating with the acetabulum of the os coxa forming a ball-and-socket joint. A ball-and-socket joint is a freely moveable synovial joint.

1. Ilium   4. Head of femur
2. Acetabulum   5. Pubis
3. Ball-and-socket joint   6. Symphysis pubis

Figure 7.3 An intervertebral joint between vertebral bodies is a type of cartilaginous joint. The intervertebral discs of intervertebral joints can be readily seen in a lateral MR image of the lower back.

1. Body of second lumbar vertebra (L2)
2. Intervertebral disc

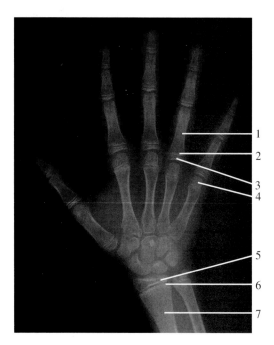

Figure 7.4 A synchondrosis is a type of cartilaginous joint located within a long bone at both the proximal and the distal epiphyseal plates as seen in a radiograph of a child's hand. Mitotic activity at a synchondrotic joint is responsible for linear (length) bone growth.

1. Diaphysis of phalanyx
2. Epiphyseal plate
3. Proximal epiphysis of phalanx
4. Distal epiphysis of metacarpal bone
5. Distal epiphysis of radius
6. Epiphyseal plate
7. Diaphysis of radius

Figure 7.5 The side-to-side articulation of the ulna and radius forms a syndesmosis that is tightly bound by an interosseous ligament (not seen in radiograph). A syndesmosis is a type of fibrous joint that permits slight movement.

1. Carpal bones	3. Radius
2. Syndesmosis	4. Ulna

Figure 7.6 Radiographs of the (a) wrist and hand and (b) the ankle and foot. Several kinds of synovial joints (freely moveable joints) are contained in these regions.

1. Hinge joints — metacarpophalangeal joints of second digit (index finger)
2. Condyloid joint — metacarpophalangeal joint of the second digit (index finger)
3. Saddle joint — carpometacarpal joint of the pollex (thumb)
4. Gliding joint — intercarpal joint between capitate and scaphoid bones
5. Condyloid joint — radiocarpal joint
6. Condyloid joint at ankle
7. Gliding joints between tarsal bones
8. Hinge joints between phalanges of digits

Figure 7.7 A radiograph of the elbow region depicting two types of synovial joints. The humeroulnar joint is a hinge joint that permits movement along a single plane. The proximal radioulnar and humeroradial joints are both pivot joints that permit rotational movement.

1. Medial epicondyle of humerus
2. Olecranon
3. Humeroulnar joint
4. Proximal radioulnar joint
5. Ulna
6. Humerus
7. Olecranon fossa
8. Lateral epicondyle of humerus
9. Humeroradial joint
10. Head of radius
11. Radial tuberosity
12. Radius

Figure 7.8 The ball-and-socket joint seen in this radiograph of the hip is a synovial joint. The coxal joint (hip joint) is formed by the head of the femur articulating with the acetabulum of the os coxa.

1. Ilium
2. Acetabulum
3. Ball-and-socket joint
4. Head of femur
5. Greater trochanter of femur
6. Pubis
7. Ischium
8. Obturator foramen
9. Femur

1. Head of humerus
2. Ball-and-socket joint
3. Humerus
4. Glenoid fossa
5. Scapula

Figure 7.9 The ball-and-socket joint seen in this radiograph of the shoulder is a synovial joint. The humeral joint (shoulder joint) is formed by the head of the humerus articulating with the glenoid fossa of the scapula.

(a) Flexion at right shoulder, elbow, and knee joints. Extension at left shoulder, elbow, and knee joints.

(b) Maximal flexion at each of the principal body joints.

(d) Adduction at the shoulder and hip joints. Also, adduction at the joints of the fingers and toes.

(c) Rotation at the joints of the neck; elevation at the shoulder joint; and flexion at the elbow and wrist joints.

(e) Abduction at the shoulder and hip joints. Also, abduction at the metacarpophalangeal joints at the base of the digits.

(f) Rotation at joints of the vertebral column.

(g) Lateral bending at joints of the vertebral column.

(h) Flexion at the joints of the vertebral column.

(i) Extension at the joints of the vertebral column.

(j) Flexion at the shoulder, hip, and knee joints on left side of body; extension at the elbow and wrist joints.

(k) Extension at the shoulder and hip joints on left side of body; plantar flexion at the left ankle joint.

Figure 7.10 A photographic summary of joint movements (a-k).

(a)

Flexion and extension at the cervical intervertebral joints of the neck.

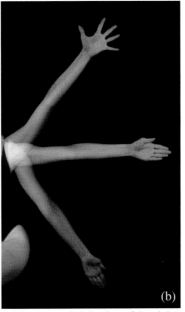

(b)

Adduction and abduction of the right arm at the shoulder joint and fingers at the metacarpophalangeal joints.

(c)

Flexion and extension at the left elbow joint.

(d)

Abduction and adduction of the hand at the wrist.

(e)

Flexion and extension at the hip and knee joints.

(f)

Flexion and extension at the knee joint and plantar flexion and dorsiflexion at the ankle joint.

Figure 7.11 A visualization of angular movements permitted at synovial joints (a-f).

Figue 7.12 Rotation of the head at the cervical vertebrae–principally at the atlantoaxial joint.

Figure 7.13 Rotation of the antebrachium (forearm) at the proximal radioulnar joint--an example of pronation and supination.

Figure 7.12 Structure of the elbow joint as seen in a radiograph.

1. Articular cavity of elbow joint
2. Head of radius
3. Neck of radius
4. Radial tuberosity
5. Radius
6. Humerus
7. Olecranon
8. Ulna

Figure 7.13 Structure of the humeral joint (shoulder joint) as seen in a MR image.

1. Glenoid labrum
2. Articular cavity of shoulder joint
3. Head of humerus

Figure 7.14 Structure of the wrist joint as seen in a MR image.

1. Heads of second and third metacarpal bones
2. Distal carpal bones (trapezoid and capitate bones)
3. Proximal carpal bones (scaphoid and lunate bones)
4. Distal epiphysis of radius

Figure 7.15 Structure of the coxal joint (hip joint) as seen in a MR image.

1. Acetabulum
2. Articular cavity of hip joint
3. Head of femur
4. Greater trochanter of femur
5. Body of femur

Figure 7.16 A sagittal MR image of tibiofemoral joint (knee joint).

1. Tendon of quadriceps muscle
2. Patella
3. Patellar ligament
4. Lateral meniscus, anterior horn
5. Head of tibia
6. Femur
7. Gastrocnemius muscle
8. Lateral meniscus, posterior horn
9. Synovial fluid
10. Articular cartilage
11. Head of fibula

Figure 7.17 A posterior MR image of tibiofemoral joint (knee joint).

1. Insertion of posterior cruciate ligament
2. Medial condyle
3. Medial meniscus
4. Tibial collateral ligament
5. Tibia
6. Lateral condyle of femur
7. Lateral meniscus
8. Synovial fluid
9. Intercondylar eminences

Figure 7.18 A sagittal MR image of the talocrual (ankle) and tarsal joints of the foot.

1. Tibia
2. Tendo calcaneus
3. Talus
4. Calcaneus
5. Navicular bone

Figure 7.19 A radiograph of the hip showing a joint prosthesis.

Muscular System

8

Most skeletal muscles span joints and are attached to a bone at both ends by a *tendon*. Certain tendons, especially at the wrist and ankle, are enclosed individually by *tendon sheaths*. A group of tendons in these areas is also covered by a *retinaculum*. The *origin* of a muscle is the more stationary attachment, and the *insertion* is the move moveable attachment (fig. 8.1). *Synergistic muscles* contract together. *Antagonistic muscles* perform in opposition to a synergistic group of muscles.

Fascia is a fibrous connective tissue that covers muscle and attaches to the skin. *Superficial fascia* secures the skin to the underlying muscle. *Deep fascia* is an inward extension of the superficial fascia and surrounds adjacent muscles compartmentalizing and binding them into functional groups.

A skeletal muscle fiber is an elongated, multinucleated, striated cell (fig. 8.2). Each fiber is surrounded by a cell membrane, called a *sarcolemma*, and the cytoplasm within the cell is called *sarcoplasm*. Each fiber contains many parallel, thread-like structures called *myofibrils*. Each myofibril is composed of smaller strands called *myofilaments* that contain the contractile proteins, *actin* and *myosin*. The regular spatial organization of the contractile proteins within the myofibrils forms the cross-banding striations characteristic of skeletal muscle. A network of membranous channels, called the *sarcoplasmic reticulum*, extends throughout the cytoplasm.

The myofibrils of a skeletal muscle fiber are arranged into compartments called *sarcomeres*. The dark and light striations of the sarcomere are due to the arrangement of the thick (myosin) and thin (actin) filaments. The dark bands are called *A bands* and the lighter bands, containing only actin filaments, are called *I bands*. The outer regions of the A bands contain actin and myosin; however, the lighter central regions *(H zones)* of the A bands contain only myosin. The I bands are bisected by dark *Z lines* where the actin filaments of adjacent sarcomeres join. Muscle fiber contraction results from the interaction of contractile proteins (actin and myosin myofilaments) in which the length of the sarcomeres is reduced.

Figure 8.1 Skeletomuscular relationship.

Figure 8.2 Structure of a muscle fiber.

Occipitalis m.
Splenius capitis m.
Trapezius m.
Deltoid m.
Long head of triceps brachii m.
Lateral head of triceps brachii m.
Medial head of triceps brachii m.
Brachioradialis m.
Extensor carpi ulnaris m.
Flexor carpi ulnaris m.
Extensor digitorum m.
Extensor retinaculum
Dorsal interosseus m.
Vastus lateralis m.
Long head of biceps femoris m.
Semitendinosus m.
Short head of biceps femoris m.
Popliteal fossa
Medial head of gastrocnemius m.
Lateral head of gastrocnemius m.
Soleus m.
Tendo calcaneus

Sternocleidomastoid m.
Infraspinatus m.
Teres minor m.
Teres major m.
Latissimus dorsi m.
Anconeus m.
External abdominal oblique m.
Gluteus medius m.
Gluteus maximus m.
Gracilis m.
Adductor magnus m.
Semimembranosus m.
Popliteus m.

Figure 8.4 A posterior view of human musculature (m=muscle).

Frontalis m.
Corrugator m.
Zygomaticus m.
Masseter m.
Depressor anguli oris m.
Trapezius m.
Deltoid m.
Pectoralis major m.
Brachialis m.
Biceps brachii m.
Rectus abdominis m.
Brachioradialis m.
Flexor carpi radialis m.
Palmaris longus m.
Gluteus medius m.
Iliopoas m.
Pectineus m.
Adductor longus m.
Rectus femoris m.
Vastus lateralis m.
Vastus medialis m.
Peroneus longus m.
Tibialis anterior m.
Extensor digitorum longus m.
Peroneus brevis m.

Orbicularis oculi m.
Orbicularis oris m.
Sternocleidomastoid m.
Coracobrachialis m.
Latissimus dorsi m.
Serratus anterior m.
Rectus sheath
Umbilicus
External abdominal oblique m.
Tensor fasciae latae m.
Abductor pollicis brevis m.
Abductor digiti minimi m.
Sartorius m.
Gracilis m.
Tendon of rectus femoris m.
Patella
Patellar ligament
Gastrocnemius m.
Soleus m.

Figure 8.3 An anterior view of human musculature (m=muscle).

Figure 8.5 An anterolateral view of the trunk.

1. Clavicle
2. Pectoralis major m. (reflected)
3. Pectoralis minor m.
4. External intercostal m.
5. Serratus anterior m.
6. External abdominal oblique m.
7. Rectus abdominis m.
8. Ribs
9. Internal intercostal m.
10. Sternum
11. Costal cartilages
12. Tendinous inscriptions of rectus abdominis m.

Figure 8.6 An anterior view of the trunk.

1. Platysma m.
2. Deltoid m.
3. Cephalic veins
4. External abdominal oblique m. (aponeurosis removed)
5. Tendinous inscriptions of rectus abdominis m.
6. Rectus abdominis m.
7. Internal abdominal oblique m.
8. Inguinal ligament
9. Pyramidalis m.
10. Sternocleidomastoid m.
11. Clavicle
12. Pectoralis major m.
13. Sternum
14. Xiphoid process
15. External abdominal oblique m.
16. Umbilicus
17. Aponeurosis of external abdominal oblique m.

Figure 8.7 A posterolateral view of the trunk.

1. Trapezius m.
2. Triangle of ausculation
3. Latissimus dorsi m.
4. Vertebral column (spinous processes)
5. Infraspinatus m.
6. Deltoid m.
7. Teres minor m.
8. Teres major m.
9. Serratus anterior m.
10. Rib
11. External abdominal oblique m.
12. Iliac crest
13. Gluteus medius m.
14. Gluteus maximus m.

Figure 8.8 An anterior view of the neck muscles.

1. Platysma m.
2. Digastric m.
3. Mylohyoid m.
4. Hyoid bone
5. Manubrium of sternum
6. Sternocleidomastoid m.
7. Thyrohyoid m.
8. Omohyoid m.
   (superior belly)
9. Thyroid cartilage
10. Sternohyoid m.
11. Clavicle
12. Rib
13. Body of sternum

Figure 8.9 An anterolateral view of the trunk.

1. Subclavius m.
2. Deltoid m.
3. Pectoralis major m. (clavicular part)
4. Cephalic vein
5. Pectoralis major m.
6. Serratus anterior m.
7. External abdominal oblique m.
8. Clavicle
9. Sternum

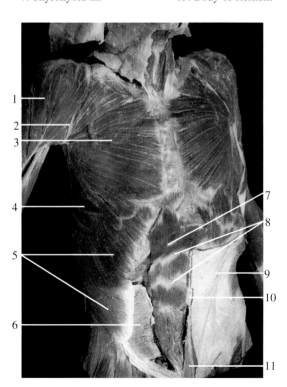

Figure 8.10 An anterolateral view of the trunk.

1. Deltoid m.
2. Cephalic vein
3. Pectoralis major m.
4. Serratus anterior m.
5. External abdominal oblique m.
6. Internal abdominal oblique m.
7. Rectus abdominis m.
8. Tendinous inscriptions
9. Aponeurosis of external abdominal oblique m.
10. Umbilicus
11. Pyramidalis m.

Figure 8.11 The deep muscles of the back and right side.

1. Splenius capitis m.
2. Levator scapulae m.
3. Rhomboideus minor m.
4. Rhomboideus major m.
5. Spinalis thoracis m.
6. Longissimus thoracis m.
7. Iliocostalis thoracis m.
8. Serratus posterior inferior m.
9. Iliac crest
10. Spine of the scapula
11. Supraspinatus m.
12. Deltoid m.
13. Teres minor m.
14. Infraspinatus m.
15. Teres major m.
16. Scapula (medial border)
17. Latissimus dorsi m. (reflected)
18. Serratus anterior m.
19. External intercostal m.
20. Rib
21. External abdominal oblique m.

Figure 8.12 The superficial muscles of the gluteal region.

1. Iliac crest
2. Gluteus medius m.
3. Gluteus maximus m.
4. Semitendinosus m.
5. Vastus lateralis m.
6. Biceps femoris m. (long head)

Figure 8.13 The deep structures of the gluteal region.

1. Iliac crest
2. Gluteus minimus m.
3. Piriformis m.
4. Superior gemellus m.
5. Obturator internus m.
6. Inferior gemellus m.
7. Quadratus femoris m.
8. Ischial tuberosity
9. Semimembranosus m.
10. Gluteus medius m. (reflected)
11. Gluteus maximus m. (reflected)
12. Sciatic nerve
13. Semitendinosus m.
14. Biceps femoris m. (long head)

Figure 8.14 The medial brachium and superficial flexors of the right antebrachium.

1. Triceps brachii m. (long head)
2. Biceps brachii m. (short head)
3. Triceps brachii m. (medial head)
4. Flexor carpi radialis m.
5. Palmaris longus m.
6. Superficial digital flexor m.
7. Flexor carpi ulnaris m.

Figure 8.15 An anterior view of the superficial muscles of the right forearm.

1. Flexor carpi radialis m.
2. Palmaris longus m.
3. Superficial digital flexor m.
4. Flexor carpi ulnaris m.

Figure 8.16 An anterior view of the deep muscles of the right forearm.

1. Pronator teres m.
2. Flexor pollicus longus m.
3. Pronator quadratus m.
4. Median nerve
5. Deep digital flexor m.

Figure 8.17 A posterior view of the superficial muscles of the right forearm.

1. Triceps brachii m. (medial head)
2. Extensor carpi radialis longus m.
3. Extensor digitorum m.
4. Extensor carpi minimi m.
5. Extensor carpi carpi ulnaris m.
6. Brachialis m.
7. Biceps brachii m. (long head)
8. Brachioradialis m.
9. Extensor carpi radialis brevis m.
10. Abductor pollicis longus m.
11. Extensor pollicus brevis m.
12. Radius
13. Extensor retinaculum
14. Tendon of extensor pollicus longus m.
15. Dorsal interosseous mm.

Figure 8.18 A posterior view of deep muscles of the left forearm.

1. Brachioradialis m.
2. Extensor carpi radialis longus m.
3. Extensor carpi radialis brevis m.
4. Abductor pollicus longus m.
5. Extensor pollicis brevis m.
6. Extensor pollicis longus m.
7. Anconeus m.
8. Supinator m.
9. Extensor indicis m.
10. Ulna

Figure 8.19 A posterior view of the right thigh.

1. Semimembranosus m.
2. Gracilis m.
3. Gluteus maximus m.
4. Semitendinosus m.
5. Biceps femoris (long head)
6. Iliotibial tract
7. Vastus lateralis m.
8. Common peroneal nerve
9. Tibial nerve
10. Gastrocnemius m. (lateral and medial heads)

Figure 8.20 The right popliteal fossa and the surrounding structures.

1. Semimembranosus m.
2. Semitendinosus m
3. Plantaris m.
4. Gastrocnemius m. (reflected)
5. Biceps femoris m.
6. Common peroneal nerve
7. Popliteus m.
8. Soleus m.

Figure 8.21 An anterior view of the right thigh.

1. Inguinal ligament
2.. Iliopsoas m.
3. Tensor faciae latae m.
4. Deep femoral artery
5. Rectus femoris m.
6. Iliotibial tract
7. Vastus lateralis m.
8. Patella
9. Pectineus m.
10. Femoral artery
11. Adductor longus m.
12. Sartorius m.
13. Gracilis m.
14. Vastus medialis m.

Figure 8.22 An anteromedial view of the muscles of the right leg.

1. Extensor digitorum longus m.
2. Tuberosity of the tibia
3. Tibialis anterior m.
4. Tibia
5. Tendon of tibialis anterior m.
6. Tendon of extensor hallucis longus m.
7. Superior extensor retinaculum

Figure 8.23 An anterior view of the dorsum of the right foot.

1. Lateral malleolus
2. Extensor digitorum brevis m.
3. Tendons of extensor digitorum longus m.
4. Tendons of extensor digitorum brevis m.
5. Tibia
6. Superior extensor retinaculum
7. Medial malleolus
8. Tendon of tibialis anterior m.
9. Tendon of extensor hallucis longus m.
10. Tendon of extensor hallucis brevis m.

Nervous System

The nervous system is anatomically divided into the *central nervous system (CNS)*, which includes the *brain* and *spinal cord*, and the *peripheral nervous system (PNS)*, which includes the *cranial nerves*, arising from the brain, and the *spinal nerves*, arising from the spinal cord (fig. 9.1). The *autonomic nervous system (ANS)* is a functionally distinct division of the nervous system devoted to regulation of involuntary activities in the body. The ANS is made up of specific portions of the CNS and PNS.

The brain and spinal cord are the centers for integration and coordination of information. Conveyed as *nerve impulses*, information to and from the brain and spinal cord travels through nerves. *Nerves* are similar to electrical conducting wires. Nerve impulses are sent from the brain in the form of electrical signals along *motor nerves* to the receiving organs, which then translate the signal into some specific function. For example, the motor impulses conducted from the brain to the muscles of the forearm that serve the hand cause the fingers to move as the muscles are contracted. *Sensory nerves* conduct nerve impulses in the opposite direction—from the receptor site to the CNS. For example, a pinprick on the skin produces a sensory impulse along a sensory nerve that the brain interprets as a painful sensation.

Neurons and *neuroglia* are the two cell types that make up nervous tissue. Neurons are specialized to respond to physical and chemical stimuli, conduct impulses, and release specific chemical regulators, called *neurotransmitters*. Although neurons vary considerably in size and shape, they have three principal components: a *cell body*, *dendrites*, and an *axon* (fig. 9.3). In a typical neuron connection, the axon of one neuron *synapses* (joins) on the cell body or dendrites of a neighboring neuron. Axons vary in length from a few millimeters in the CNS to over a meter in the PNS. Long axons are generally *myelinated* with *neurolemmocytes (Schwann cells)* in the PNS, and many of the short axons are myelinated with *oligodendrocytes* in the CNS. *Neurofibril nodes (nodes of Ranvier)* are segments in the *myelin sheath*. The end of the axon at the synapse is called the *axon terminal*.

Figure 9.2 A photomicrograph of a neuron.
1. Cytoplasmic extensions
2. Nucleus
3. Cell body of neuron

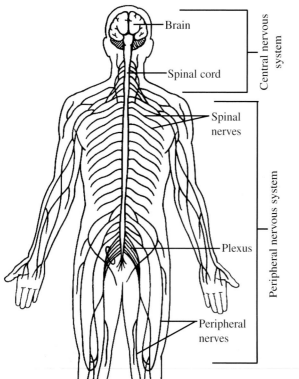

Figure 9.1 Divisions of the nervous system.

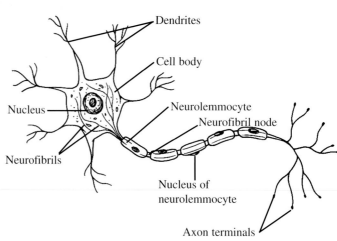

Figure 9.3 Structure of a myelinated neuron.

70

Figure 9.4 A sectioned cranium exposing the meninges and cerebrum.

1. Skull
2. Cerebrum
3. Dura mater

Figure 9.5 The relationship of the meninges to the skull and the cerebrum.

1. Skin of scalp
2. Galea aponeurotica
3. Bone of cranium
4. Dura mater
5. Arachnoid mater

6. Subarachnoid space
7. Blood vessel
8. Pia mater
9. Cerebral cortex
10. White matter of brain

Figure 9.6 A lateral view of the brain.

1. Gyri
2. Sulci
3. Frontal lobe of cerebrum
4. Lateral sulcus
5. Olfactory cerebral cortex
6. Temporal lobe of cerebrum
7. Primary sensory cerebral cortex
8. Central sulcus
9. Primary motor cerebral cortex
10. Parietal lobe of cerebrum
11. Occipital lobe of cerebrum
12. Visual cerebral cortex
13. Auditory cerebral cortex
14. Cerebellum
15. Medulla oblongata

Figure 9.7 An inferior view of the brain with the eyes and part of the meninges still intact.

1. Longitudinal cerebral fissure
2. Muscles of the eye
3. Temporal lobe of cerebrum
4. Pituitary gland
5. Pons
6. Medulla oblongata
7. Cerebellum
8. Spinal cord

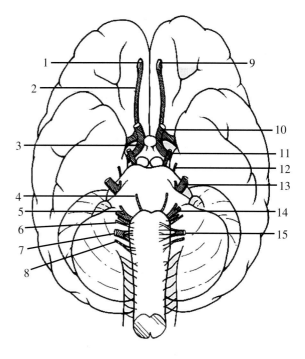

Figure 9.8 A diagram of the inferior of the brain showing the cranial nerves.

1. Olfactory bulb
2. Olfactory tract
3. Optic tract
4. Abducens nerve
5. Facial nerve
6. Glossopharyngeal nerve
7. Vagus nerve
8. Accessory nerve
9. Olfactory nerve
10. Optic nerve
11. Oculomotor nerve
12. Trochlear nerve
13. Trigeminal nerve
14. Vestibulocochlear nerve
15. Hypoglossal nerve

Figure 9.9 Cranial nerves and blood supply to the brain.

1. Internal carotid artery
2. Cerebral arterial circle (circle of Willis)
3. Trigeminal nerve
4. Vestibulocochlear nerve
5. Abducens nerves
6. Hypoglossal nerve
7. Olfactory tract
8. Optic nerve
9. Oculomotor nerve
10. Trochlear nerve
11. Trigeminal nerve
12. Facial nerve
13. Glossopharyngeal nerve
14. Vagus nerve
15. Vertebral artery

Figure 9.10 A sagittal view of the brain.

1. Truncus of corpus callosum
2. Crus of fornix
3. Third ventricle
4. Posterior commissure
5. Splenium of corpus callosum
6. Pineal body
7. Inferior colliculus
8. Arbor vitae of cerebellum
9. Vermis of cerebellum
10. Choroid plexus of fourth ventricle
11. Tonsilla of cerebellum
12. Medulla oblongata
13. Septum pellucidum (cut)
14. Intraventricular foramen
15. Genu of corpus callosum
16. Anterior commissure
17. Hypothalmus
18. Optic chiasma
19. Oculomotor nerve
20. Cerebral peduncle
21. Midbrain
22. Pons
23. Fourth ventricle
24. Pyramid

Figure 9.11 Transaxial section of the skull and brain.

1. First ventricle
2. Second ventricle
3. Falx cerebri (septum of dura mater)
4. Cingulate gyrus
5. Genu of corpus callosum
6. Caudate nucleus
7. Choroid plexus
8. Splenium of corpus callosum
9. Falx cerebri (septum of dura mater)

Figure 9.12 Coronal MRI brain scan.

1. Cerebrospinal fluid
2. Longitudinal cerebral fissure
3. Third ventricle
4. Cerebellum
5. Skull
6. Dura mater
7. Cerebral cortex
8. Cerebral medulla
9. Lateral ventricle
10. Fourth ventricle

Figure 9.13 MRI sagittal section through the skull.

1. Genu of corpus callosum
2. Frontal sinus
3. Pituitary gland
4. Ethmoidal sinus
5. Sphenoidal sinus
6. Tegmentum (midbrain)
7. Pons
8. Truncus of corpus callosum
9. Fornix
10. Thalamus
11. Splenium of corpus callosum
12. Pineal gland
13. Superior and inferior colliculi
14. Fourth ventricle
15. Cerebellum
16. Medulla oblongata

Figure 9.14 MRI transaxial section through the brain.

1. Frontal sinus
2. Frontal horn of lateral ventricle
3. Body of fornix
4. Posterior horn of lateral ventricle
5. Splenium of corpus callosum
6. Genu of corpus callosum
7. Head of caudate nucleus
8. External capsule
9. Thalamus
10. Choroid plexus

Figure 9.15 MRI transaxial section showing visual pathways.

1. Superior colliculi
2. Lateral ventricle
3. Third ventricle
4. Optic nerve
5. Optic chiasma
6. Optic tract
7. Lateral geniculate body
8. Calcarine tracts (optic radiation)

Figure 9.16 A MRI coronal section through the thalamus.

1. Superior sagittal sinus
2. Lateral ventricle
3. Lateral fissure
4. Corpus callosum
5. Thalamus
6. Third ventricle

Figure 9.18 Anterior surface of the brain and spinal cord with meninges removed.

1. Cervical enlargement
2. Lumbar enlargement
3. Sciatic nerve
4. Sympathetic ganglia
5. Cauda equina

Figure 9.17 A posterior view of the lower spinal cord.

1. Spinal cord
2. Dura mater (cut)
3. Posterior (dorsal) root of spinal nerve
4. Cauda equina
5. Filum terminale

Figure 9.19 Sheep brain, dorsal view. (a) photograph; (b) diagram.

Figure 9.20 Sheep brain, ventral view. (a) photograph; (b) diagram.

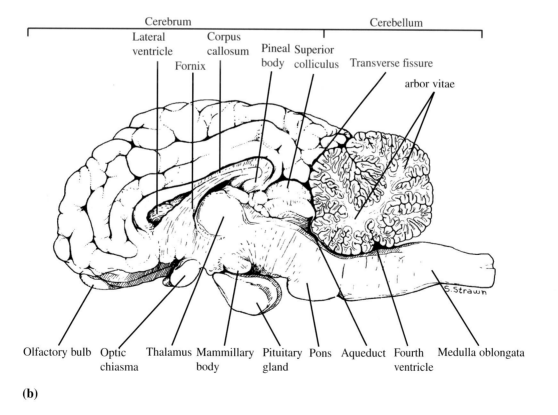

Figure 9.21 Sheep brain, sagittal view. (a) photograph; (b) diagram.

Endocrine System

10

The endocrine system works closely with the nervous system to regulate and integrate body processes and maintain homeostasis. The nervous system regulates body activities through the action of electrochemical impulses that are transmitted by means of neurons, resulting in rapid, but usually brief responses. By contrast, the endocrine system is composed of glands (fig. 10.1) scattered throughout the body that release chemical substances called *hormones* into the bloodstream. These hormones dissipate in the blood and travel throughout the entire body to act on *target tissues*, where they have a slow but relatively long-lasting effect. Neurological responses are measured in milliseconds, but hormonal action requires seconds or days to elicit a response. Some hormones may have an effect that lasts for minutes and others, for weeks or months.

The endocrine system and nervous system are closely coordinated in autonomically controlling the functions of the body. The *pituitary gland*, located in the brain, regulates the activity of most other endocrine glands. Located immediately between the pituitary and the rest of the brain is the *hypothalamus* The hypothalamus serves as an intermediate between the nervous centers of the brain and the pituitary gland, correlating the activity of the two systems. Furthermore, certain hormones may stimulate or inhibit the activities of the nervous system.

Other organs of the endocrine system include the *thyroid gland* and *parathyroid glands*, located in the neck. The *adrenal glands* and *pancreas* are located in the abdominal region. The *ovaries* of the female are located in the pelvic cavity, whereas the *testes* of the male are located in the scrotum. Even the *placenta* serves as an endocrine organ for the developing fetus and has some hormonal influence upon the pregnant woman.

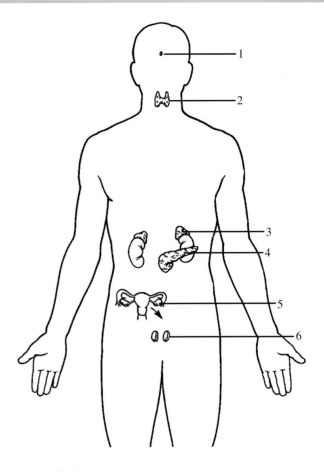

Figure 10.1 Principal endocrine glands.

1. Pituitary gland	4. Pancreas
2. Thyroids and parathyroids	5. Ovary
3. Adrenal (suprarenal) gland	6. Testis

Figure 10.2 Pituitary gland.　　　　7X

1. Pars intermedia (adenohypophysis)
2. Pars distalis (adenohypophysis)
3. Pars nervosa (neurohypophysis)

Figure 10.3 Pituitary gland.　　　　40X

1. Pars intermedia
2. Pars distalis
3. Pars nervosa

Figure 10.4 Pars distalis of the pituitary gland. 400X
1. Basophil
2. Chromophobe
3. Acidophil

Figure 10.5 Pineal gland. 100X
1. Brain sand
2. Pinealocytes

400X

Figure 10.6 Thyroid gland.
1. Follicle cells
2. C cells
3. Colloid within follicle

75X

Figure 10.7 Parathyroid gland.
1. Chief cells
2. Chief cells
3. Cluster of oxyphil cells

Figure 10.8 The adrenal (suprarenal) gland.
1. Adrenal gland
2. Inferior suprarenal artery
3. Kidney

Figure 10.9 Adrenal gland.   7X
1. Adrenal cortex
2. Adrenal medulla
3. Adrenal cortex
4. Blood vessel

Figure 10.10 Adrenal gland.   75X
1. Capsule
2. Zona glomerulosa (adrenal cortex)
3. Zona fasiculata (adrenal cortex)
4. Zona reticularis (adrenal cortex)
5. Adrenal medulla

Figure 10.11 Pancreatic islet (islet of Langerhan's). 75X
1. Pancreatic islet (endocrine pancreas)
2. Acini (exocrine pancreas)

Sensory Organs

11

The nervous and endocrine systems convey information from the brain to all parts of the body to enable a person to interact with both the external and internal environments and to maintain homeostasis. The sense organs, in contrast, convey information from the outside world (and inside world of the body) back to the brain. This includes a wide range of information such as temperature, brightness, sound, flavor, and balance.

The sense organs are actually extensions of the nervous system that allow us to autonomically respond or conscientiously perceive our internal and external environments. A stimulus excites a sense organ which then transduces the stimulus to an electrical (nerve) impulse. Sensory nerves transmit the impulse (sensation) to the brain to be perceived and acted upon. Ultimately, it is the brain which actually feels, sees, hears, tastes, and smells.

The *eyes* are the organs of visual sense. The eyes refract (bend) and focus the incoming light waves onto the sensitive *photoreceptors (rods and cones)* at the back of each eye. Nerve impulses from the stimulated photoreceptors are conveyed along visual pathways to the occipital lobes of the cerebrum, where visual sensations are perceived.

The eyeball consists of the fibrous tunic, which is divided into the *sclera* and *cornea*; the vascular tunic, which consists of the *choroid*, the *ciliary body*, and the *iris*; and the internal tunic, or *retina*, which consists of an outer pigmented layer and an inner nervous layer. The eye contains an anterior cavity between the lens and the retina. The anterior cavity is subdivided into an anterior chamber in front of the iris and a posterior chamber behind the iris. *Aqeous humor* fills both of these chambers. The posterior cavity (also called the vitreous chamber) contains vitreous humor.

The *ear* is the organ of hearing and equilibrium (balance). It contains receptors that respond to movements of the head and receptors that convert sound waves into nerve impulses. Impulses from both receptor types are transmitted through the vestibulocochlear (VIII) cranial nerve to the brain for interpretation.

The ear consists of the three principal regions: the *outer ear*, the *middle ear*, and the *inner ear*. The outer ear consists of the *auricle* and the *external auditory canal*. The middle ear contains the auditory ossicles (*malleus, incus,* and *stapes*). The inner ear contains the *spiral organ* (organ of Corti) in the *cochlea* for hearing, and the *semicircular canals* for equilibrium.

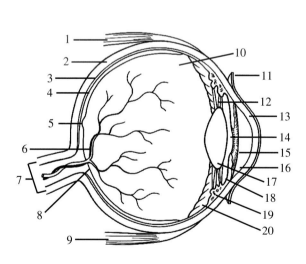

Figure 11.1 Structure of the eye.

1. Superior rectus m.	11. Conjuctiva
2. Sclera	12. Suspensory ligament
3. Choroid	13. Cornea
4. Retina	14. Pupil
5. Fovea centralis	15. Iris
6. Central vessels	16. Anterior chamber
7. Optic nerve	17. Lens
8. Optic disc	18. Posterior chamber
9. Inferior rectus m.	19. Ciliary body
10. Vitreous chamber	20. Ora serrata

Figure 11.2 Structure of the ear.

1. Helix	8. Facial nerve
2. Auricle	9. Vestibulocochlear nerve
3. External auditory canal	10. Cochlea
4. Earlobe	11. Vestibular (oval) window
5. Outer ear	12. Auditory ossicles
6. Middle ear	13. Auditory tube
7. Inner ear	14. Tympanic membrane

Figure 11.3 The surface anatomy of the eye.

1. Eyebrow
2. Superior eyelid (palpebra)
3. Palpebra commisure
4. Lateral canthus
5. Sclera
6. Eyelashes
7. Pupil
8. Iris
9. Lacrimal caruncle
10. Medial commisure
11. Conjuctiva
12. Inferior eyelid (palpebra)

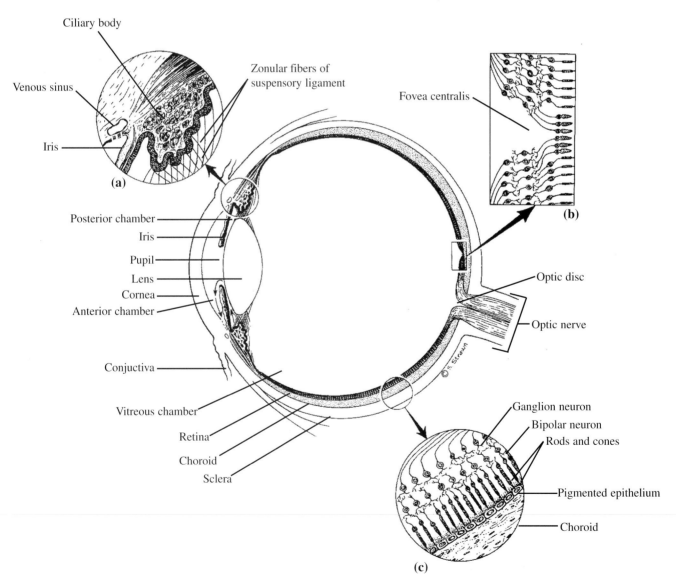

Figure 11.4 Structure of the eye. (a) The ciliary body, (b) fovea centralis, and (c) retina.

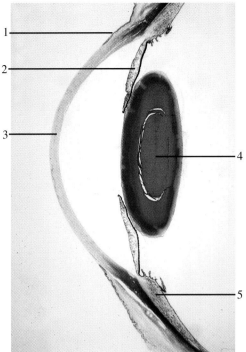

Figure 11.5 Anterior portion of the eye.   7X

1. Conjunctivia   4. Lens
2. Iris   5. Ciliary body
3. Cornea

Figure 11.6 Retina.   250X

1. Retina   3. Choroid
2. Rods and cones   4. Sclera

Figure 11.7 The surface anatomy of the auricle.

1. Helix   5. Tragus
2. Triangular fossa   6. External auditory canal
3. Antihelix   7. Antitragus
4. Concha   8. Earlobe

Figure 11.8 Cochlea.   75X

1. Scala vestibuli   3. Cochlear duct
2. Vestibular membrane   4. Scala tympani

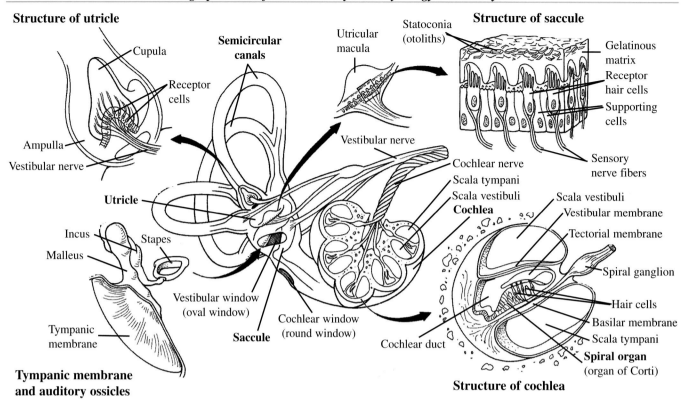

Structure of utricle

Cupula

Receptor cells

Ampulla

Vestibular nerve

Semicircular canals

Utricular macula

Utricle

Incus

Stapes

Malleus

Tympanic membrane

Tympanic membrane and auditory ossicles

Vestibular window (oval window)

Vestibular nerve

Cochlear nerve

Scala tympani

Scala vestibuli

Cochlea

Saccule

Cochlear window (round window)

Cochlear duct

Statoconia (otoliths)

Structure of saccule

Gelatinous matrix

Receptor hair cells

Supporting cells

Sensory nerve fibers

Scala vestibuli

Vestibular membrane

Tectorial membrane

Spiral ganglion

Hair cells

Basilar membrane

Scala tympani

Spiral organ (organ of Corti)

Structure of cochlea

Figure 11.9 Structures of the middle ear and inner ear. The tympanic membrane and auditory ossicles (malleus, incus, stapes) are structures of the middle ear. The vestibular organs (utricle, saccule, semicircular canals) and cochlea (containing the spiral organ) are structures of the inner ear.

Figure 11.10 Spiral organ (organ of Corti).   250X

1. Vestibular membrane
2. Cochlear duct
3. Tectorial membrane
4. Hair cells
5. Basilar membrane

Figure 11.11 Taste bud.   200X

1. Epithelium of papilla
2. Taste buds
3. Tongue muscle

Circulatory System

12

The circulatory system consists of the *blood*, *heart*, and *vessels*, each of which is essential to the life of a complex multicellular organism. Blood, a specialized connective tissue, consists of *formed elements* (erythrocytes, leukocytes, and thrombocytes) that are suspended and carried in the *plasma*. These formed elements function in transport, immunity, and blood-clotting mechanisms.

The heart is enclosed in a *pericardial sac* within the thoracic cavity. The wall of the heart consists of the *epicardium*, *myocardium*, and *endocardium*. The *right atrium* of the heart receives blood from the superior vena cava and inferior vena cava, and the *right ventricle* pumps blood into the *pulmonary trunk* to the *pulmonary arteries*. The *left atrium* receives blood from the *pulmonary veins* and pumps blood into the left ventricle. The left ventricle pumps blood into the *aorta*.

There are four heart valves that prohibit the backflow of blood: 1) The *right tricuspid valve (right atrioventricular valve)* is located between the right atrium and the right ventricle; 2) the *pulmonary semilunar valve (pulmonary valve)* is located between the right ventricle and the pulmonary trunk; 3) the *left bicuspid valve (left atrioventricular, or mitral, valve)* is located between the left atrium and the left ventricle; and, 4) the *aortic semilunar valve (aortic valve)* is located between the left ventricle and the ascending aorta.

The *systemic arteries* arise from the aorta or branches of the aorta and transport blood away from the heart to smaller vessels called *arterioles*. From arterioles, the blood enters *capillaries* where diffusion with the surrounding cells may occur. Capillaries converge forming *venules*, which in turn converge forming larger vessels called *veins*. Veins are vessels that transport blood toward the heart.

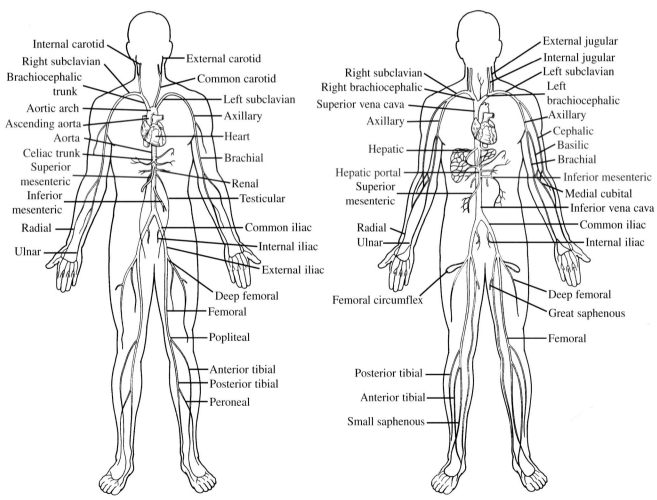

Figure 12.1 Principal arteries of the body.

Figure 12.2 Principal veins of the body.

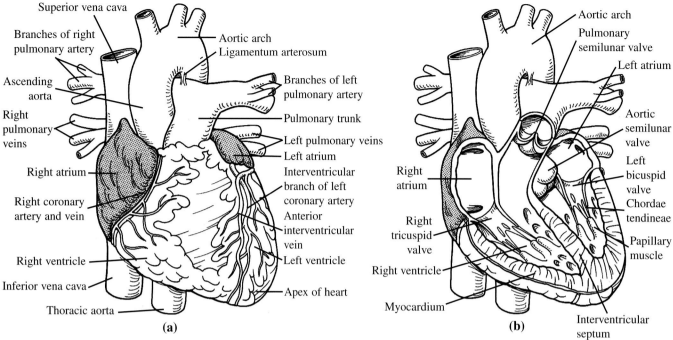

Superior vena cava
Branches of right pulmonary artery
Ascending aorta
Right pulmonary veins
Right atrium
Right coronary artery and vein
Right ventricle
Inferior vena cava
Thoracic aorta

Aortic arch
Ligamentum arterosum
Branches of left pulmonary artery
Pulmonary trunk
Left pulmonary veins
Left atrium
Interventricular branch of left coronary artery
Anterior interventricular vein
Left ventricle
Apex of heart

(a)

Aortic arch
Pulmonary semilunar valve
Left atrium
Aortic semilunar valve
Left bicuspid valve
Chordae tendineae
Papillary muscle
Interventricular septum

Right atrium
Right tricuspid valve
Right ventricle
Myocardium

(b)

Figure 12.3 The structure of the heart. (a) An anterior view and (b) an internal view.

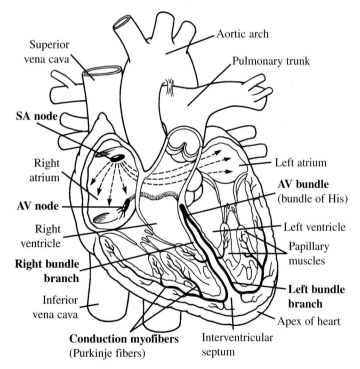

Superior vena cava
SA node
Right atrium
AV node
Right ventricle
Right bundle branch
Inferior vena cava
Conduction myofibers (Purkinje fibers)

Aortic arch
Pulmonary trunk
Left atrium
AV bundle (bundle of His)
Left ventricle
Papillary muscles
Left bundle branch
Apex of heart
Interventricular septum

Figure 12.4 Conduction system of the heart, indicated in bold type.

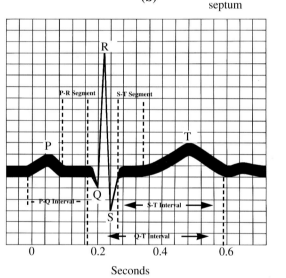

Figure 12.5 Normal electrocardiogram (ECG).

Figure 12.6 Wall of the heart.   7X

1. Epicardium   3. Endocardium
2. Myocardium

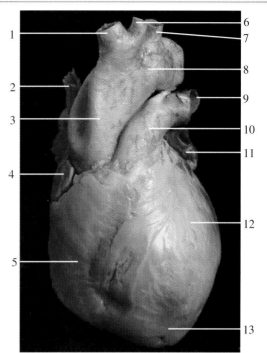

Figure 12.7 The position of the heart within the pericardium.

1. Mediastinum   4. Diaphragm
2. Right lung   5. Liver
3. Pericardium   6. Left lung

Figure 12.8 An anterior view of the heart and great vessels.

1. Brachiocephalic trunk   8. Aortic arch
2. Superior vena cava   9. Pulmonary artery
3. Ascending aorta   10. Pulmonary trunk
4. Right atrium   11. Left atrium
5. Right ventricle   12. Left ventricle
6. Left common carotid artery 13. Apex of heart
7. Left subclavian artery

Figure 12.9 A posterior view of the heart.

1. Aortic arch   5. Right ventricle
2. Pulmonary arteries   6. Ascending aorta
3. Right atrium   7. Left ventricle
4. Inferior vena cava

Figure 12.10 The internal structure of the heart.

1. Right atrium   7. Aortic semilunar valve
2. Tricuspid valve   8. Bicuspid valve
3. Right ventricle   9. Myocardium
4. Interventricular septum 10. Papillary muscle
5. Trabeculae carneae   11. Left ventricle
6. Ascending aorta

(a)

Position the heart so the ventral surface faces you. Notice the thicker ventricular walls, especially the left ventricle.

1. Aortic arch
2. Pulmonary trunk
3. Left ventricle

(b)

Insert the scissors into the superior vena cava. The cut should expose the interior of the right atrium. Notice the right tricuspid (atrioventricular) valve.

1. Right atrium
2. Superior vena cava

(c)

Continue the incision through the right ventricle to the apex of the heart. Observe the structure of the valve.

1. Right tricuspid valve
2. Right ventricle

(d)

Begin the next incision in the left atrium. This time, continue through both the atrium and the ventricle.

1. Left ventricle
2. Left atrium

(e)

1. Left bicuspid (mitral) valve.
2. Left ventricle

Expose the left ventricle and atrium. Notice the difference between the right and left ventricles, especially the thicker muscular wall of the left ventricle.

Figure 12.11 Sheep heart dissection.

Figure 12.12 An angiogram showing the aortic arch and its branches.

1. External carotid arteries
2. Right common carotid artery
3. Right vertebral artery
4. Right subclavian artery
5. Brachiocephalic trunk
6. Aortic arch
7. Internal carotid arteries
8. Left vertebral artery
9. Left subclavian artery
10. Left common carotid artery

Figure 12.13 An angiogram showing the branches of the common carotid and external carotid arteries.

1. Meningeal arteries
2. Occipital artery
3. Internal carotid artery
4. Common carotid artery
5. Internal carotid artery to cerebral arterial circle (circle of Willis)
6. Maxillary artery
7. External carotid artery
8. Facial artery
9. Superior thyroid artery

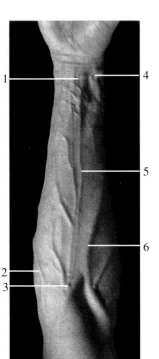

Figure 12.14 Surface anatomy identifying the superficial veins of the forearm.

1. Tendon of palmaris longus m.
2. Basilic vein
3. Median cubital vein
4. Site for palpation of radial artery
5. Median antebrachial vein
6. Cephalic vein

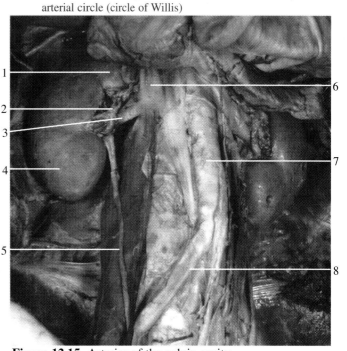

Figure 12.15 Arteries of the pelvic cavity.

1. Suprarenal gland
2. Renal artery
3. Renal vein
4. Right kidney
5. Ureter
6. Inferior vena cava
7. Abdominal aorta
8. Right common iliac artery

Figure 12.16 Arteries of the pelvic cavity.

1. Right common iliac artery
2. External iliac artery
3. Inguinal ligament
4. Femoral artery
5. Abdominal aorta
6. Left common iliac artery

Figure 12.17 An angiogram of the common iliac arteries and their branches.

1. Common iliac artery
2. External iliac artery
3. Femoral artery
4. Deep femoral artery
5. Lumbar vertebra
6. Internal iliac artery
7. Gluteal arteries
8. Obturator artery
9. Symphysis pubis
10. Lateral circumflex femoral artery

Figure 12.18 An angiogram of the arteries of the right thigh.

1. Deep femoral artery
2. Lateral circumflex femoral artery
3. Medial femoral circumflex artery
4. Femoral artery
5. Femur

Figure 12.19 Wall of elastic artery.   200X

1. Tunica adventitia
2. Elastic laminae (in tunica media)
3. Tunica intima

Figure 12.20 Artery and vein.   40X
1. Artery
2. Vein

Figure 12.21 Wall of muscular artery.   75X
1. Tunica adventitia   3. Internal elastic membrane
2. Tunica media   4. Endothelial cells

Photo courtesy of Scott Miller

Figure 12.22 SEM photomicrograph of a capillary.

1. Nucleus
2. Endocytic vesicles
3. Endothelial cell
4. Lumen of capillary
5. Basal lamina

Figure 12.23 Arteriole, capillary, and venule.

1. Capillary
2. Arteriole
3. Venule

200X

Figure 12.24 Wall of large vein.   200X

1. Tunica adventita
2. Longitudinally oriented smooth muscle
3. Tunica media
4. Tunica intima

Figure 12.25 Electron micrograph of blood cells in the lumen of a blood vessel.

1. Erythrocytes   2. Leukocytes

Photo courtesy of Clifford E. Keeney

Figure 12.26 Electron micrograph of a blood clot.

1. Erythrocytes   3. Leukocyte
2. Thrombocytes   4. Fibrin strand

Figure 12.27 Types of leukocytes.   200X

(a) Neutrophil   (d) Monocyte (1);
(b) Basophils   lymphocyte (2)
(c) Eosinophil

Lymphatic System

13

The lymphatic system is closely interrelated to the circulatory system. The functions of the lymphatic system are basically fourfold: 1) it transports excess interstitial (tissue) fluid, which was initially formed as a blood filtrate, back to the bloodstream; 2) it maintains homeostasis around body cells by providing a constantly moist intracellular environment, which assists movements of materials into and out of cells; 3) it serves as the route by which absorbed fat from the small intestine is transported to the blood; and 4) it helps provide immunological defenses against disease-causing agents.

Lymph capillaries drain tissue fluid, which is formed from blood plasma; when this fluid enters lymph capillaries, it is called *lymph*. Lymph is returned to the venous system via two large lymph ducts—the *thoracic duct* and the *right lymphatic duct* (fig. 13.1). On the way to these drainage ducts, lymph filters through *lymph nodes*, which contain phagocytic cells and germinal centers that produce lymphocytes. The *spleen* and *thymus* are considered lymphoid organs because they also produce lymphocytes.

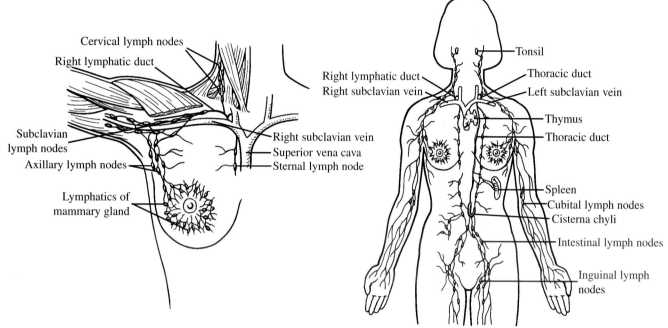

Cervical lymph nodes
Right lymphatic duct
Subclavian lymph nodes
Axillary lymph nodes
Lymphatics of mammary gland
Right subclavian vein
Superior vena cava
Sternal lymph node

Tonsil
Right lymphatic duct
Right subclavian vein
Thoracic duct
Left subclavian vein
Thymus
Thoracic duct
Spleen
Cubital lymph nodes
Cisterna chyli
Intestinal lymph nodes
Inguinal lymph nodes

Figure 13.1 Structure of the lymphatic system.

Figure 13.2 Thymus.   7X
1. Cortex
2. Medulla

Figure 13.3 Thymic medulla.   300X
1. Hassall's corpuscle

Figure 13. 4　The thymus within a fetus, during the third trimester of development.

1. Thymus
2. Lung
3. Heart
4. Liver
5. Umbilical cord

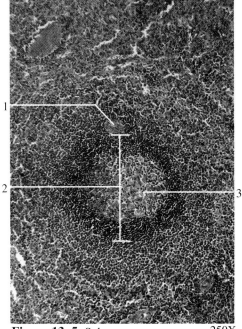

Figure 13. 5　Spleen.　　　　250X

1. Central artery
2. Splenic nodule
3. Germinal center

Figure 13. 6　The spleen and pancreas.

1. Spleen
2. Splenic artery
3. Splenic vein
4. Pancreas
5. Pancreatic duct

Figure 13. 7 Palatine tonsil.   40X
1. Oral mucosa
2. Lymphatic nodule
3. Germinal centers

Figure 13.8 Palatine tonsils that have been removed in a tonsillectomy and sectioned by a pathologist. Chronic tonsillitis generally requires a tonsillectomy.

Figure 13.9 Lymph node.   40X
1. Medulla of lymph node   3. Lymphatic nodule
2. Capsule   4. Cortex of lymph node

Figure 13.10 A lymph node.
1. Lymph node
2. Vein
3. Lymphatic vessels
4. Muscle

Respiratory System

14

The respiratory system is made up of organs and structures that function together to bring gases in contact with the blood of the circulatory system. This system consists of the *nasal cavity, pharynx, larynx*, and *trachea*, and the *bronchi, bronchioles*, and *alveoli* within the lungs (fig. 14.2). The functions of the respiratory system are gas exchange, sound production, assistance in abdominal compression, and coughing and sneezing.

The nasal cavity has a bony and cartilaginous support. The ciliated, mucous lining of the upper respiratory tract warms, moistens, and cleanses inspired air. The *paranasal sinuses* are found in the maxillary, frontal, sphenoid, and ethmoid bones. The *pharynx* is a funnel-shaped passageway that connects the oral and nasal cavities with the larynx. The cartilaginous *larynx* keeps the passageway to the trachea open during breathing and closes the respiratory passageway during swallowing. It also contains the *vocal cords*. The *trachea* is a rigid tube, supported by rings of cartilage, that leads from the larynx to the *bronchial tree*. *Pulmonary alveoli* are the functional units of the lungs where gas exchange occurs; they are small, numerous, thin-walled air sacs. The right and left *lungs* are separated by the *mediastinum*; each lung is divided into *lobes* and *lobules*.

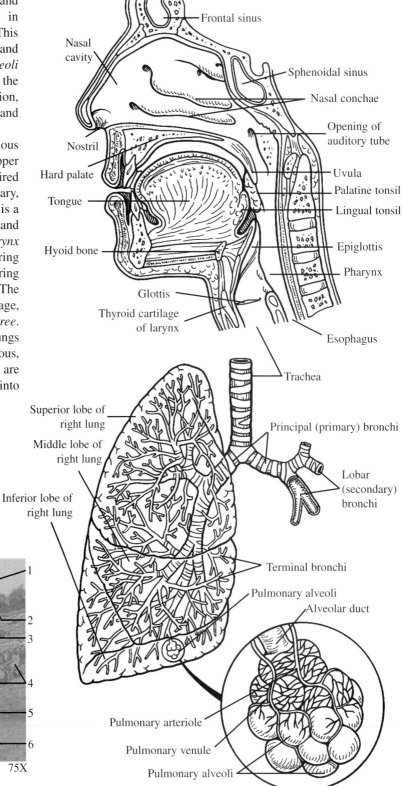

Figure 14.1 Tracheal wall.　　75X

1. Respiratory epithelium
2. Basement membrane
3. Duct of seromucous gland
4. Seromucous glands
5. Perichondrium
6. Hyaline cartilage

Figure 14.2 Structure of the respiratory system.

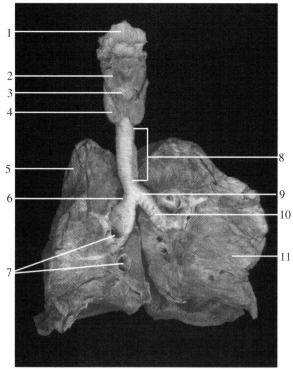

Figure 14.3 An anterior view of the larynx, trachea, and lungs.

1. Epiglottis
2. Thyroid cartilage
3. Cricoid cartilage
4. Thyroid gland
5. Right lung
6. Right principal (primary) bronchus
7. Pulmonary vessels
8. Trachea
9. Carnia
10. Left principal (primary) bronchus
11. Left lung

Figure 14.4 An anterior view of the larynx and trachea.

1. Epiglottis
2. Thyroid cartilage
3. Cricothyroid ligament
4. Thyroid gland
5. Superior lobe of right lung
6. Right principal (primary) bronchus
7. Cricoid cartilage
8. Isthmus of thyroid gland
9. Superior lobe of left lung
10. Trachea
11. Carnia
12. Left principal (primary) bronchus

Figure 14.5 An anterior view of bronchi.

1. Trachea
2. Apical segmental bronchus
3. Posterior segmental bronchus
4. Anterior segmental bronchus
5. Right principal bronchus
6. Lateral segmental bronchus
7. Medial segmental bronchus
8. Anterior basal segmental bronchus
9. Carina
10. Left principal bronchus
11. Esophagus
12. Medial basal segmental bronchus
13. Posterior basal segmental bronchus
14. Lateral segmental bronchus

Figure 14.6 Radiograph of the thorax.

1. Thoracic vertebra
2. Right lung
3. Rib
4. Image of right breast
5. Diaphragm/liver
6. Clavicle
7. Left lung
8. Mediastinum
9. Heart
10. Diaphragm/stomach

Figure 14.7 Bronchiole.   75X

1. Pulmonary arteriole   3. Alveoli
2. Bronchiole

Figure 14.8 Electron micrograph of the lining of the trachea.

1. Cilia
2. Goblet cell

300X

Figure 14.9 Terminal   200X
bronchiole.

1. Terminal bronchiole
2. Respiratory bronchiole
3. Alveolar duct

Figure 14.10 Respiratory alveoli.

1. Capillary in alveolar wall
2. Macrophages
3. Type II pneumocytes

Digestive System

15

The digestive system consists of a *gastrointestinal tract* (GI tract) and *accessory digestive organs*. Most of the food we eat is not suitable for cellular utilization until it is mechanically and chemically reduced to forms that can be absorbed through the intestinal wall and transported to the cells by the blood or lymph. Ingested food is not technically in the body until it is absorbed and, in fact, a large portion of consumed food is not digested at all but rather passes through as fecal material.

The functions of the principal regions and organs of the digestive system are presented in table 15.1. The digestive system is diagrammed in figure 15.1.

Table 15.1 Regions and Structures of the Digestive System

	Region or Structure	*Function*
Gastrointestinal tract	Oral cavity	Ingests food; receives saliva and initiates digestion of carbohydrates; mastication (chewing); forms bolus (food mass); deglutition (swallowing)
	Pharynx	Receives bolus from oral cavity and passes it to esophagus
	Esophagus	Transports bolus to stomach by peristalsis
	Stomach	Receives bolus from esophagus; forms chyme (paste-like food) initiates digestion of proteins; moves chyme into duodenum
	Small intestine	Receives chyme form stomach, along with secretions from liver and pancreas; chemically and mechanically breaks down chyme; absorbs nutrients; transports wastes to large intestine
	Large intestine	Receives undigested wastes from small intestine; absorbs water and electrolytes; forms and stores feces, and expels them through defecation
Accessory digestive organs	Teeth	Mechanically pulverize food
	Tongue	Manipulate food and assists in swallowing
	Salivary glands	Secrete saliva which aids in formation of bolus; initiates digestion of carbohydrates
	Liver	Production of bile; storage of iron and copper; conversion of glucose to glycogen and storage of glycogen; synthesis of certain vitamins; production of urea; synthesis of fibrinogen and prothrombin used for clotting of blood; phagocytosis of foreign material in blood; detoxifies harmful substances in body; storage of blood cells; hemopoiesis in fetus and newborn
	Gallbladder	Concentration and storage of bile necessary for emulsification of fats
	Pancreas	Production and secretion of pancreatic juice containing digestive enzymes; production and secretion of the hormones insulin and glucagon

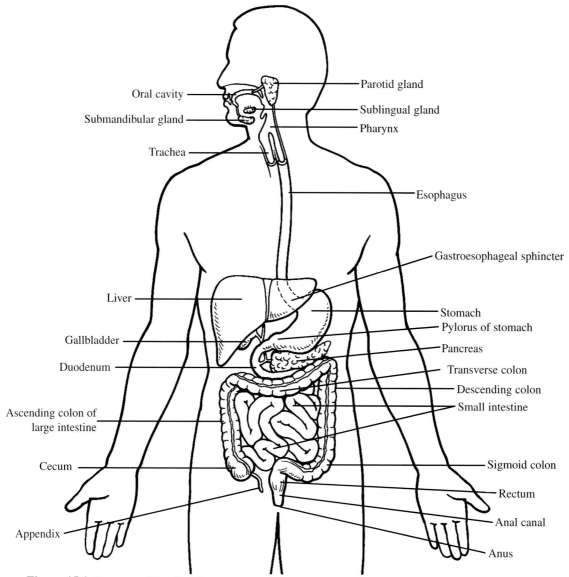

Figure 15.1 Structure of the digestive system.

Figure 15.2 Oral region, lips, and teeth.

1. Medial incisor	6. Medial incisor
2. Lateral incisor	7. Philtrum
3. Canine	8. Canine
4. Canine	9. Inferior lip
5. Lateral incisor	10. Mentolabial sulcus

Figure 15.3 Structures of the oral cavity with the mouth open and the tongue elevated.

1. Tongue
2. Lingual frenulum
3. Opening of submandibular duct

Figure 15.4   Filiform and fungiform papillae.   75X

1. Filiform papillae   2. Fungiform papilla

Figure 15.5   Vallate papilla   40X

1. Vallate papilla   2. Taste buds

Figure 15.6   Developing tooth.   40X

1. Ameoblasts   4. Odontoblasts
2. Enamel   5. Pulp
3. Dentin

Figure 15.7   Mature tooth.   7X

1. Dentin (enamel has been   3. Gingiva
    dissolved away)   4. Alveolar bone
2. Pulp

Figure 15.8 Acini of salivary tissue.   250X

1. Serous acinus   3. Mucous acinus
2. Serous demilune on   4. Serous demilune on
    mucous acinus   mucous acinus

Figure 15.9 Striated duct.   250X

1. Mucous acini   3. Striated duct
2. Serous acinus

Figure 15.10 Sublingual gland (mostly mucous, 100X
some serous).

1. Mucous acini   2. Serous demilune

Figure 15.11 Submandibular gland (about one-half   100X
mucous and one-half serous).

1. Serous acinus   2. Mucous acinus

Figure 15.12 Parotid gland (purely serous).   250X

1. Serous acini   3. Striated duct
2. Excretory duct

Figure 15.13 Cross section of esophagus.   10X

1. Mucosa   3. Muscularis
2. Submucosa   4. Lumen

Figure 15.14 Wall of esophagus.   30X

1. Inner circular layer (muscularis externa)
2. Outer longitudinal layer (muscularis externa)
3. Epithelium
4. Smooth muscle
5. Skeletal muscle

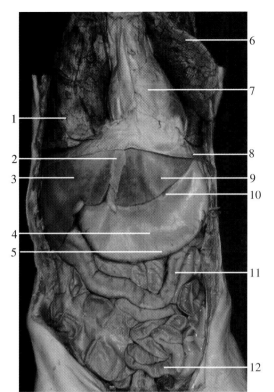

Figure 15.15 An anterior aspect of the trunk.

1. Right lung
2. Falciform ligament
3. Right lobe of liver
4. Body of stomach
5. Greater curvature of stomach
6. Left lung (reflected)
7. Pericardium
8. Diaphragm
9. Left lobe of liver
10. Lesser curvature of stomach
11. Transverse colon
12. Small intestine

Figure 15.16 An anterior view of the stomach and liver.

1. Left lobe of liver
2. Falciform ligament
3. Celiac trunk (covered by hepatoduodenal ligament)
4. Gallbladder
5. Right lobe of liver
6. Pylorus of stomach
7. Transverse colon
8. Fundus of stomach
9. Lesser omentum
10. Gastric vein (traversing through omentum
11. Body of stomach

Figure 15.17 A CT scanogram of the trunk.

1. Right lung within pleural cavity
2. Thoracic vertebra
3. Liver
4. Small intestine
5. Ascending colon
6. Heart
7. Diaphragm
8. Stomach
9. Descending colon
10. Ilium

Figure 15.18 Thoracic and abdominal viscera.

1. Lungs
2. Heart (surrounded by pericardial fat)
3. Diaphragm
4. Liver
5. Cecum
6. Small intestine

Figure 15.19 Upper abdominal organs. The greater omentum has been removed and the stomach reflected.

1. Right lobe of liver
2. Duodenum
3. Taeniae coli
4. Right colic (hepatic) flexure
5. Mesentery
6. Greater curvature of stomach
7. Lesser omentum
8. Pancreas
9. Transverse colon
10. Left colic (splenic) flexure
11. Small intestine

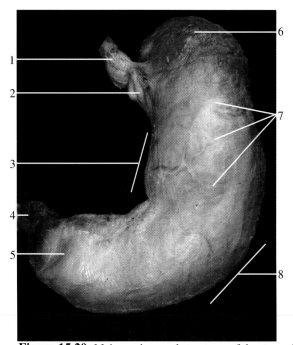

Figure 15.20 Major regions and structures of the stomach.

1. Esophagus
2. Cardiac portion of stomach
3. Lesser curvature of stomach
4. Duodenum
5. Pylorus of stomach
6. Fundus of stomach
7. Body of stomach
8. Greater curvature of stomach

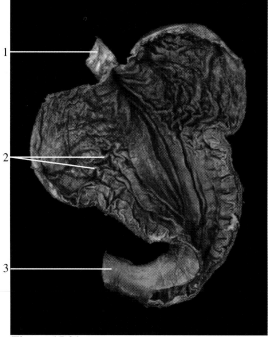

Figure 15.21 Interior aspect of the stomach.

1. Esophagus
2. Gastric folds (rugae)
3. Duodenum

Figure 15.22   Junction of esophagus   10X
and stomach.

1. Epithelium of esophagus
2. Abrupt change in epithelium
3. Epithelium of stomach

Figure 15.23   Wall of stomach.   10X

1. Mucosa   3. Muscularis externa
2. Submucosa

Figure 15.24   Gastric glands.   300X

1. Parietal cells
2. Chief cells

Figure 15.25   Mucosa of stomach body.   30X

1. Mucosa   3. Gastric glands
2. Gastric pits   4. Muscularis mucosae

Figure 15.26   Mucosa of stomach pylorus.   200X

1. Gastric pit   3. Gastric glands
2. Lamina propria

Figure 15.27 An anterior aspect of the small and large intestines with associated structures.

1. Taeniae coli
2. Appendix
3. Cecum
4. Left colic (splenic) flexure
5. Transverse colon
6. Small intestine
7. Epiploic appendages

Figure 15.28 Mesentery and small intestine.

1. Serosa (visceral peritoneum)
2. Mesentery

Figure 15.29 Sections of the small intestine.

1. Serosa (visceral peritoneum)
2. Ileum
3. Jejunum
4. Plicae circulares

Figure 15.30 Junction of pylorus and duodenum.   10X

1. Stomach epithelium   4. Pyloric sphincter
2. Duodenal epithelium
3. Duodenal (Brunner's)
   glands

Figure 15.31 Small intestine.   7X

1. Villi   3. Submucosa
2. Plicae circulares   4. Muscularis externa

Figure 15.32 Small intestine.   30X

1. Intestinal glands   4. Submucosa
2. Villus   5. Muscularis externa
3. Muscularis mucosae

Figure 15.34 Intestinal villus.   30X

1. Lamina propria   4. Simple columnar
2. Villus   epithelium
3. Central lacteal

Figure 15.33 Intestinal glands.   300X

1. Eosinophil within lamina   3. Intestinal gland
   propria   4. Paneth cells
2. Muscularis mucosae

Figure 15.35 Section of the large intestine (colon).

1. Haustrum
2. Taeniae coli
3. Epiploic appendage
4. Semilunar folds (plicae) of colon

Figure 15.36 Radiograph of the large intestine.

1. Right colon (hepatic) flexure
2. Ascending colon
3. Sigmoid colon
4. Cecum
5. Left colic (splenic) flexure
6. 12th rib
7. Transverse colon
8. Lumbar vertebra
9. Descending colon
10. Rectum
11. Hip joint

Figure 15.37 The cecum and appendix.

1. Ascending colon
2. Taenia coli
3. Cecum
4. Ileum
5. Appendix

Figure 15.38 Electron micrograph of the ileum, shown in cross section.

1. Villi
2. Lumen
3. Mucosa
4. Submucosa
5. Tunica muscularis
6. Adventitia

Figure 15.39 An inflamed appendix that has been removed in an appendectomy. The chief danger of appendicitis is the the appendix might rupture and produce peritonitis.

10X

Figure 15.40 Appendix.

1. Lymphatic nodule
2. Circular layer of muscularis externa
3. Longitudinal layer of muscularis externa
4. Serosa (peritoneum)

75X

Figure 15.41 Large intestine.

1. Glands
2. Lamina propria
3. Muscularis mucosae

Figure 15.42 An inferior view of the liver and gallbladder.

1. Gallbladder
2. Right lobe of liver
3. Cystic duct
4. Common hepatic duct
5. Common bile duct
6. Falciform ligament
7. Left lobe of liver
8. Hepatic artery
9. Caudate lobe of liver
10. Hepatic portal vein

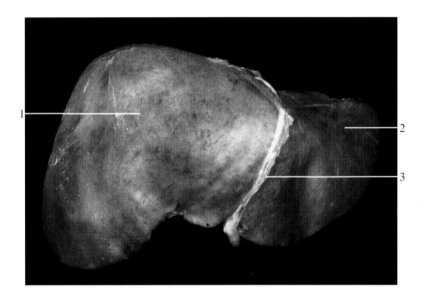

Figure 15.43 An anterior view of the liver.

1. Right lobe of liver
2. Left lobe of liver
3. Falciform ligament

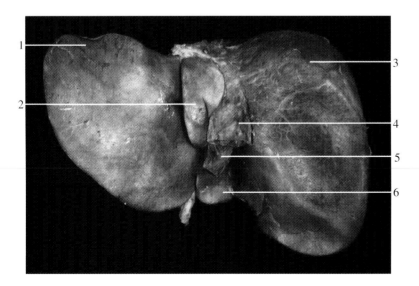

Figure 15.44 An inferior view of the liver.

1. Left lobe of liver
2. Caudate lobe of liver
3. Right lobe of liver
4. Hepatic portal vein
5. Hepatic artery
6. Quadrate lobe of liver

Figure 15.45 Hepatic lobules. 75X

1. Portal area   3. Central vein
2. Hepatic lobule

Figure 15.46 Portal area. 75X

1. Portal vein   3. Hepatic artery
2. Bile duct

1000X

Figure 15.47 Hepatic plates.

1. Sinusoid containing red blood cells
2. Kupffer cell with debris in it's cytoplasm
3. Hepatocyte nucleus
4. Plate of hepatocytes

Figure 15.48 A gallbladder that has been removed (cholecystectomy) and cut open to remove it's gallstone (biliary calculus).

1. Gallbladder
2. Gallstone

100X

Figure 15.49 Gallbladder.

1. Mucosal folds
2. Lamina propria
3. Muscularis

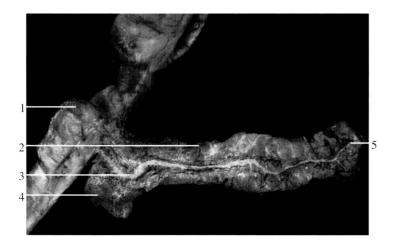

Figure 15.50 An anterior aspect of the pancreas and pancreatic duct.

1. Duodenum
2. Body of pancreas
3. Pancreatic duct
4. Head of pancreas
5. Tail of pancreas

625X

Figure 15.51 Pancreatic acini.

1. Centroacinar cells
2. Acinus

Urinary System

16

The urinary system consists of the *kidneys, ureters, urinary bladder*, and *urethra* (fig. 16.1). The urinary system: 1) removes metabolic wastes from the blood and excretes it (as urine) to the outside during micturition (the physiological aspect of urination); 2) regulates, in part, the rate of red blood cell formation by secretion of the hormone *erythropoietin*; 3) assists the regulation of blood pressure by secreting the enzyme *renin*; 4) assists the regulation of calcium by activating vitamin D; and 5) assists the regulation of the volume, composition, and pH of body fluids.

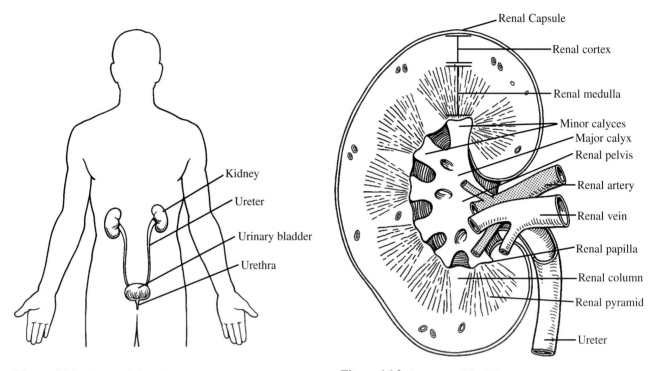

Figure 16.1 Organs of the urinary system.

Figure 16.2 Structure of the kidney.

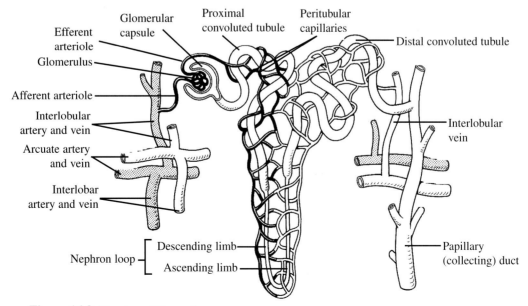

Figure 16.3 Structure of the nephron.

Figure 16.4 Kidney and ureter with overlying viscera removed.

1. Liver
2. Suprarenal gland
3. Renal artery
4. Renal vein
5. Right kidney
6. Quadratus lumborum muscle
7. Gallbladder
8. Inferior vena cava
9. Ureter

Figure 16.5 An anterior view of the right kidney.

1. Adrenal gland
2. Suprarenal artery
3. Renal artery
4. Hilum
5. Renal vein
6. Ureter

Figure 16.6 A coronal section of the left kidney.

1. Renal artery
2. Renal vein
3. Left testicular vein
4. Ureter
5. Capsule
6. Major calyx
7. Renal pelvis
8. Renal papilla
9. Renal medulla
10. Renal cortex

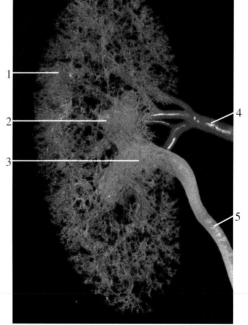

Figure 16.7 A plastic vascular cast of a kidney and ureter.

1. Arteriole network
2. Major calyx
3. Renal pelvis
4. Renal artery
5. Ureter

Figure 16.8 A transaxial CT image of abdominal viscera. Note the large renal cyst in the right kidney.

1. Superior mesenteric artery
2. Pancreas
3. Inferior vena cava
4. Renal crest
5. Body of lumbar vertebra
6. Right kidney
7. Vertebral foramen
8. Subcutaneous fat
9. Small intestine
10. Abdominal portion of aorta
11. Left ureter
12. Renal pelvis
13. Renal cortex
14. Rib
15. Skin

Figure 16.9 Kidney.   10X

1. Renal cortex
2. Renal medulla
3. Medullary rays
4. Cortical labyrinth

Figure 16.10 Medullary ray.   100X

1. Proximal tubule (of straight portion)
2. Distal tubule (of straight portion)
3. Collecting tubule
4. Glomerulus
5. Medullary ray

Figure 16.11 Electron micrograph of the renal cortex.

1. Interlobular artery
2. Afferent arteriole
3. Glomerulus
4. Efferent arteriole
5. Peritubular capillaries
6. Glomeruli

Figure 16.12 Renal corpuscle.   250X

1. Glomerular capsule
2. Urinary space
3. Glomerulus
4. Macula densa of distal tubule
5. Renal corpuscle

Figure 16.13 Cortical labyrinth   75X

1. Proximal convoluted tubule
2. Distal convoluted tubule

Figure 16.14 Renal papilla.   100X

1. Renal papilla   3. Transitional epithelium
2. Minor calyx

Figure 16.15 Wall of urinary bladder.   10X

1. Transitional epithelium   3. Muscularis
2. Lamina propria

Figure 16.16 Ureter.   15X

1. Adventitia   3. Mucosa
2. Muscularis   4. Lumen

Figure 16.17 A pyelogram of the urinary system.

1. 12th rib
2. Major calyx
3. Ureter
4. Sacroiliac joint
5. Renal pelvis
6. 3rd lumbar vertebra
7. Urinary bladder

Figure 16.18 A radiograph showing a renal calculus (kidney stone) in the renal pelvis of the right kidney.

1. Renal calculus
2. Major calyx of left kidney
3. Renal pelvis of kidney
4. Left ureter
5. Urinary bladder

Figure 16.19 Renal calculi (stones) from the urinary bladders of two different patients. Renal calculi vary considerably in appearance and size and may develop in the renal pelvis of the kidney or in the urinary bladder.

Reproductive System

17

The organs of the male and female reproductive systems produce gametes and provide the mechanism for the union of these sex cells during the process of coitus (sexual intercourse). It is through sexual reproduction that individuals of a species are propagated, each having a genetic diversity inherited from both parents.

The male reproductive system consists of the *testes* (in the *scrotum*), excretory glands (*seminal vesicles, prostate, and bulbourethral glands*), and *penis*. The functions of the male reproductive system are to secrete sex hormones, produce spermatozoa, and ejaculate semen (spermatozoa and additives) into the vagina of the female.

The female reproductive system consists of the *ovaries, uterine tubes, vagina, external genitalia,* and *mammary glands*. The functions of the female reproductive system are to secrete sex hormones, produce ova, receive ejaculated semen from the erect penis of the male, nourish the developing embryo and fetus, deliver the baby, and nurse the infant once it is born.

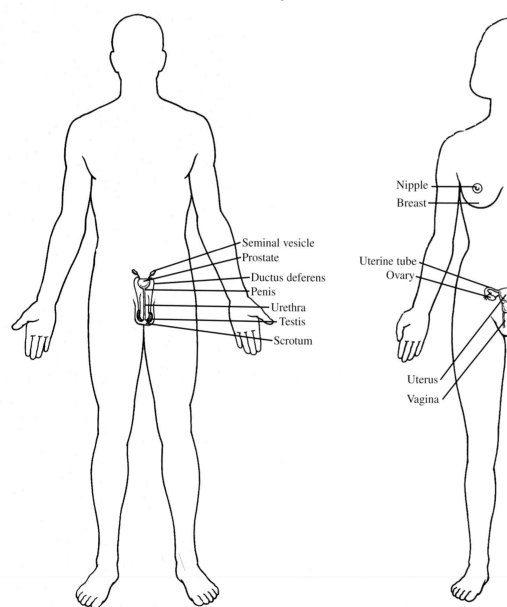

Figure 17.1 Organs of the male reproductive system.

Figure 17.2 Organs of the female reproductive system.

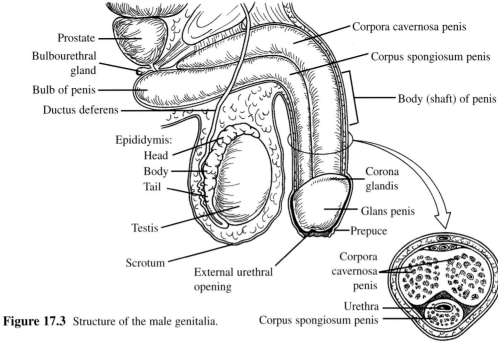

Figure 17.3 Structure of the male genitalia.

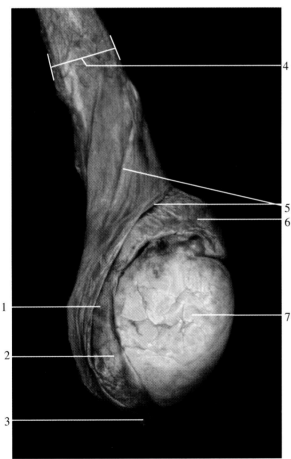

Figure 17.5 Testis and associated structures.

1. Body of epididymis
2. Tail of epididymis
3. Gubernaculum
4. Spermatic cord
5. Spermatic fascia
6. Head of epididymis
7. Testis

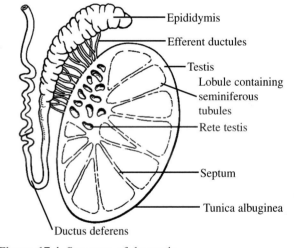

Figure 17.4 Structure of the testis.

Figure 17.6 Testis.　　　　　　　　　　　　7X
1. Tunic albuginea
2. Tubules of rete testis
3. Mediastinum
4. Seminiferous tubules

Figure 17.7 Electron micrograph of a seminiferous tubule.
1. Spermatozoa   3. Spermatogonia
2. Spermatids

Figure 17.8 Testis.   200X
1. Interstitial (Leydig) cells
2. Seminiferous tubule

Figure 17.9 Wall of seminiferous tubule.   600X
1. Sustentacular (Sertoli) cells   4. Primary spermatocytes
2. Spermatogonia   5. Boundary of seminiferous
3. Spermatids   tubule

Figure 17.10 Epididymis.   200X
1. Sperm in lumen   3. Stereocilia
2. Pseudostratified columnar
   epithelium

Figure 17.11 Efferent ductules.   75X
1. Duct of epididymis
2. Efferent ductules

Figure 17.12 Spermatic cord.   10X
1. Cremaster muscle   3. Ductus deferens
2. Veins of the pampinform   4. Testicular artery
   plexus   5. Cremaster muscle

Figure 17.13 Ductus deferens.   250X

1. Stereocilia
2. Pseudostratified columnar epithelium
3. Lamina propria
4. Muscularis

Figure 17.14 Seminal vesicle.   30X

1. Fibromuscular stroma
2. Mucosal folds
3. Lumen

Figure 17.15 Prostate.   75X

1. Prostate concretions
2. Glandular acini
3. Fibromuscular stroma

Figure 17.16 Structure of the penis.

1. Corpora cavernosa   5. Body of penis
2. Corpus spongiosum   6. Skin of penis
3. Crus of penis   7. Corona glandis
4. Bulb of penis   8. Glans penis

1X

Figure 17.17 Penis.

1. Corpora cavernosa
2. Tunica albuginea
3. Urethra
4. Corpus spongiosum

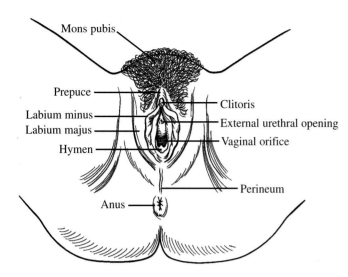

Figure 17.18 Female external genitalia.

Figure 17.19 External genitalia and internal reproductive organs of the female reproductive system..

Figure 17.20 External genitalia and vagina.

1. Mons pubis
2. Clitoris
3. Urethral opening
4. Labium minus
5. Labium majus
6. Vagina (dissected open)
7. Fornix
8. Uterine ostium
9. Cervix of uterus

Figure 17.21 Vagina.   75X

1. Nonkeratinized stratified squamous epithelium
2. Lamina propria

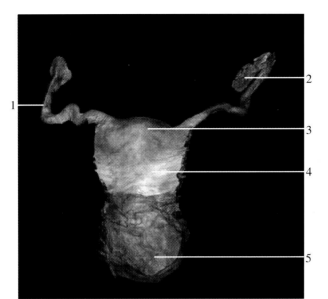

Figure 17.22 A posterior view of the uterus and uterine tubes.

1. Uterine tube
2. Fimbriae
3. Fundus of uterus
4. Body of uterus
5. Cervix of uterus

Figure 17.23 Ampulla of uterine tube.   10X

1. Lumen
2. Mucosal folds
3. Muscularis
4. Serosa

Figure 17. 24 Uterus: proliferative phase.   30X

1. Endometrial stroma
2. Lamina functionalis
   (endometrium)
3. Lamina basalis
   (endometrium)
4. Uterine glands
5. Myometrium

Figure 17. 25 Uterus: secretory phase.   30X

1. Endometrium
2. Myometrium
3. Endometrial stroma
4. Uterine glands (now
   saw-toothed)

Figure 17. 26 Uterus: menstrual   30X
phase.

1. Disintegrating lamina   3. Myometrium
   functionalis   4. Blood in stroma
2. Lamina basalis (still
   intact)

Figure 17.27 Structure of the ovary.

1. Corona radiata   8. Germinal epithelium
2. Secondary oocyte   9. Primary follicles
3. Ovulation   10. Ovarian vessels
4. Follicular fluid within antrum 11. Corpus albicans
5. Cumulus oophorus   12. Ovarian medulla
6. Oocyte   13. Ovarian cortex
7. Follicular cells   14. Corpus luteum

30X

Figure 17.28 Ovary.

1. Primordial follicles
2. Primary follicles
3. Atretic follicle
4. Tunical albuginea

200X

Figure 17.29 Secondary follicle.

1. Granulosa cells
2. Antrum
3. Zona pellucida
4. Oocyte
5. Theca interna
6. Theca externa

Figure 17. 30 Corpus luteum.   100X

1. Corpus luteum
2. Wall of corpus luteum
3. Former follicular antrum

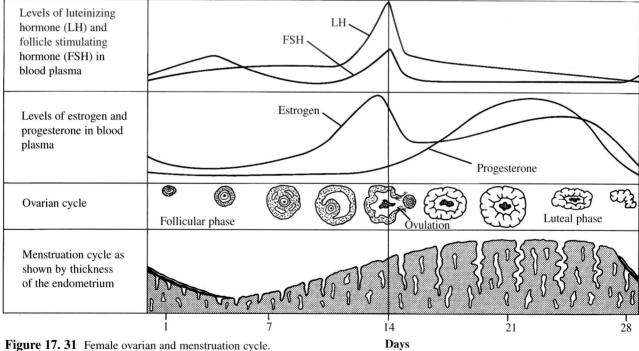

Figure 17. 31 Female ovarian and menstruation cycle.

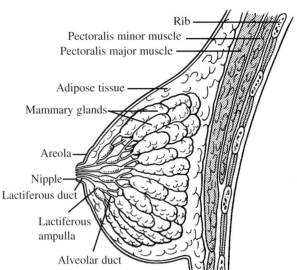

Figure 17. 32 Surface anatomy of the female breast.

1. Pectoralis major muscle
2. Axilla
3. Lateral process of breast
4. Areola
5. Nipple
6. Breast (containing mammary glands)

Figure 17. 33 Mammary gland.

40X

Figure 17. 34 Mammary glands, (non-lactating glands).

1. Interlobular duct
2. Interlobular connective tissue
3. Lobule of glandular tissue
4. Intralobular connective tissue

400X

Figure 17. 35 Mammary glands, (lactating glands).

1. Lobules of glandular tissue
2. Intralobular connective tissue tissue
3. Adipose cells

Developmental Biology

18

The period of human pregnancy, which generally requires 38 weeks, is known as *gestation. Morphogenesis* is the sequence of changes that occur in the formation of the baby's body structures. Although gestation is frequently discussed chronologically as trimesters, prenatal development is more accurately divided morphogenically into three periods based on structural changes. On this basis, the *pre-embryonic period* includes the first two weeks following fertilization, the *embryonic period* includes the following six weeks, and the *fetal period* includes the final 30 weeks.

The events of the two-week pre-embryonic period include transportation of the fertilized egg, or *zygote*, through the uterine tube, mitotic divisions, implantation, and the formation of primordial embryonic tissue (fig. 18.1). Implantation begins between the fifth and seventh day and is made possible by the secretion of enzymes that digest a portion of the endometrium of the uterus. During implantation, the *trophoblast cells* secrete *human chorionic gonadotrophin* (hCG), which prevents the breakdown of the endometrium and menstruation. The trophoblast cells also participate in the formation of the placenta.

The events of the six-week embryonic period include the differentiation of the germ layers into specific body organs and the formation of the extraembryonic membranes, including the *placenta, umbilical cord, amnion, yolk sac, allantois,* and *chorion.*

A small amount of tissue differentiation and organ development occurs during the fetal period, but for the most part fetal development is primarily limited to body growth. Labor and parturition (childbirth) are the culmination of gestation and require the action of *oxytocin* from the mother's pituitary gland, and *prostaglandins*, produced in her uterus.

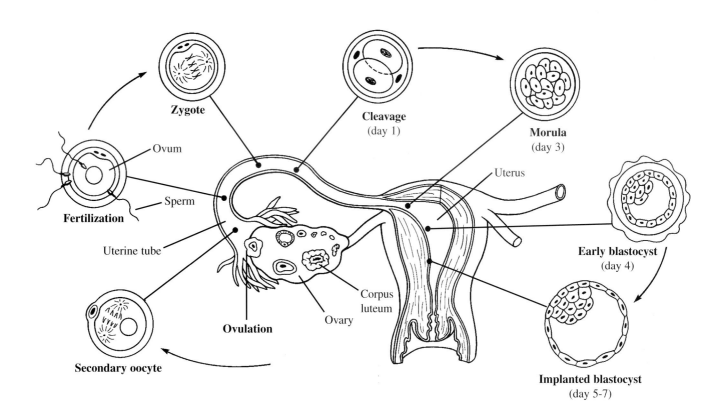

Figure 18.1 Events of ovulation, fertilization, and implantation.

Figure 18.2 A 8.1 week intrauterine pregnancy. Marks are indicated on the image to denote the crown and rump. The crown-rump length of this embryo is 18 mm. Ultrasonography, produced by a mechanical vibration of high frequency, produces a safe, high resolution of fetal structure. Most ultrasound scans are obtained on fetuses older than 12 weeks.

Figure 18.3 A human fetus at aproximately 24 weeks.

1. Umbilical cord
2. Placenta

Figure 18.4 A human fetus at aproximately 28 weeks.

1. Thymus
2. Lung
3. Heart
4. Diaphragm
5. Falciform ligament
6. Liver
7. Small intestine
8. Umbilical cord

Figure 18.5 Thoracic and abdominal viscera of a human fetus at approximately 28 weeks.

1. Pectoralis major m.
2. Pericardium (cut)
3. Pericardial cavity
4. Falciform ligament
5. External abdominal oblique m.
6. Umbilicus
7. Umbilical vein
8. Thymus
9. Lung
10. Pleural cavity
11. Heart
12. Thoracic wall
13. Diaphragm
14. Liver
15. Peritoneal cavity
16. Abdominal wall
17. Small intestine
18. Umbilical cord

Figure 18.6 Parturition, or childbirth.

Figure 18.7 Nursing.

Figure 18.8 Umbilical cord.   7X

1. Mucoid connective tissue   2. Umbilical arteries
   (Wharton's jelly)   3. Umbilical vein

Figure 18.9 Placenta.   300X

1. Nuclei of syncytiotrophoblast   3. Intervillus space containing maternal
2. Blood vessel containing fetal blood cells   blood cells
                                          4. Choronic villus

Cat Dissection

19

Embalmed cats purchased from biological supply houses are excellent specimens for dissecting and learning basic mammalian anatomy. Before the muscles and viscera of a cat can be studied, the specimen's skin has to be removed according to the following suggested guidelines.

1. Place the cat on a dissecting tray dorsal side up. Using a sharp scalpel, make a short, shallow incision through the skin across the nape of the neck. With your scissors, continue a dorsal midline incision forward over the skull and down the back to about two inches onto the tail. Sever the tail with bone shears or a saw and discard.

2. Make an incision around the neck and down each foreleg to the paws. Continue a circular incision around each wrist. Beginning at the base of the tail, make incisions down each of the hind legs to the ankles. Make a circular cut around each ankle.

3. Carefully remove the skin, using your fingers or a blunt probe to separate the skin from underlying connective tissue. Where it is necessary to use a scalpel, keep the cutting edge directed toward the skin away from the muscle. If your specimen is a male, make an incision around the genitalia, leaving the skin intact. If your specimen is a female, the mammary glands will appear as longitudinal, glandular masses along the ventral side of the abdomen and thorax. They should be remove with the skin.

4. After the specimen is skinned, remove the excess fat and connective tissue to expose the underlying muscles. Make certain that the muscles are separated along their natural boundaries. If a transection of a muscle is necessary, isolate the muscle from its attached connective tissue and make a clean cut across the belly of the muscle, leaving the origin and insertion intact.

5. At the end of the laboratory period, wrap your specimen in muslin cloth and store it in a tight, heavy-duty plastic bag. Wet your specimen from time to time with a preservative solution (usually 2%-3% phenol). Caution needs to be taken when using a phenol wetting solution as it is caustic and poisonous if misused or used in a concentrated form.

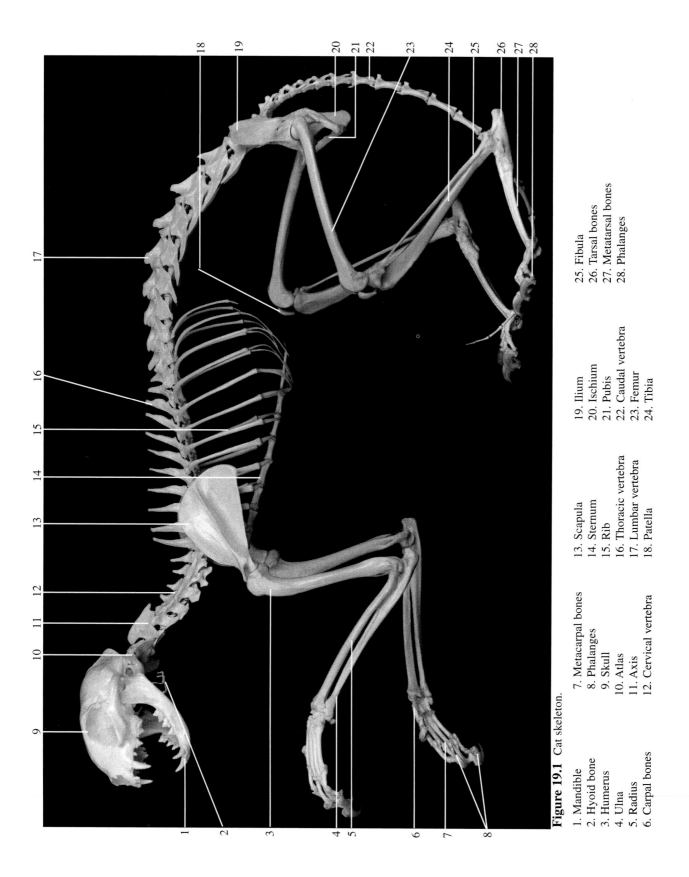

Figure 19.1 Cat skeleton.

1. Mandible
2. Hyoid bone
3. Humerus
4. Ulna
5. Radius
6. Carpal bones

7. Metacarpal bones
8. Phalanges
9. Skull
10. Atlas
11. Axis
12. Cervical vertebra

13. Scapula
14. Sternum
15. Rib
16. Thoracic vertebra
17. Lumbar vertebra
18. Patella

19. Ilium
20. Ischium
21. Pubis
22. Caudal vertebra
23. Femur
24. Tibia

25. Fibula
26. Tarsal bones
27. Metatarsal bones
28. Phalanges

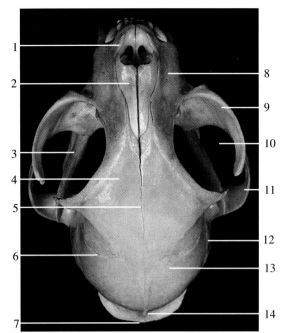

Figure 19.2 A dorsal view of a cat skull.

1. Premaxilla	9. Zygomatic
2. Nasal bone	(malar) bone
3. Mandible	10. Orbit
4. Frontal bone	11. Zygomatic arch
5. Sagittal suture	12. Temporal bone
6. Coronal suture	13. Parietal bone
7. Nuchal crest	14. Interparietal bone
8. Maxilla	

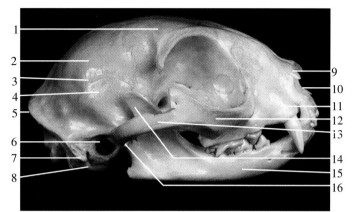

Figure 19.3 A lateral view of a cat skull.

1. Frontal bone	9. Nasal bone
2. Parietal bone	10. Premaxilla bone
3. Squamosal suture	11. Maxilla
4. Temporal bone	12. Zygomatic (malar) bone
5. Nuchal crest	13. Zygomatic arch
6. External acoustic meatus	14. Coronoid process of mandible
7. Mastoid process	15. Mandible
8. Tympanic bulla	16. Condylar process of mandible

Figure 19.4 A lateral view of the superficial muscles of the cat.

1. Clavotrapezius m.	6. Long head of triceps	11. Xiphihumeralis m.	16. Gluteus medius m.
2. Sternomastoid m.	brachii m.	12. Latissimus dorsi m.	17. Gluteus maximus m.
3. Acromiodeltoid m.	7. Acromiotrapezius m.	13. External abdominal	18. Caudofemoralis m.
4. Clavobrachialis m.	8. Spinodeltoid m.	oblique m.	19. Biceps femoris m.
5. Lateral head of triceps	9. Spinotrapezius m.	14. Lumbodorsal fascia	20. Tensor fasciae latae m.
brachii m.	10. Pectoralis minor m.	15. Sartorius m.	

Figure 19.6 A dorsal view of the cat neck and thorax.

1. Temporalis m.
2. Clavotrapezius m.
3. Acromiotrapezius m.
4. Latissimus dorsi m.
5. Supraspinatus m.
6. Rhomboideus m.
7. Serratus anterior m.

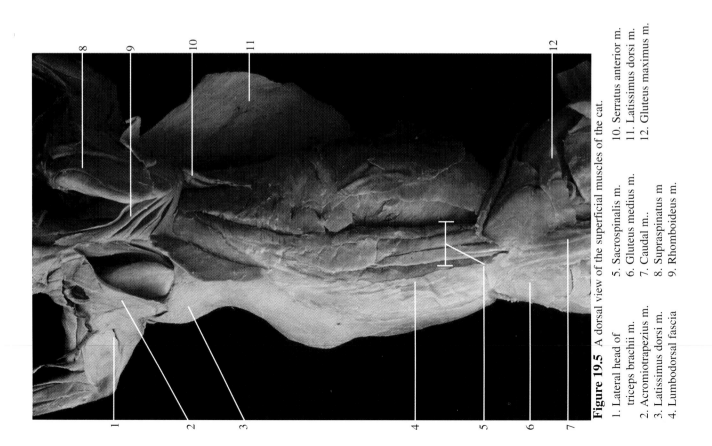

Figure 19.5 A dorsal view of the superficial muscles of the cat.

1. Lateral head of triceps brachii m.
2. Acromiotrapezius m.
3. Latissimus dorsi m.
4. Lumbodorsal fascia
5. Sacrospinalis m.
6. Gluteus medius m.
7. Caudal m..
8. Supraspinatus m
9. Rhomboideus m.
10. Serratus anterior m.
11. Latissimus dorsi m.
12. Gluteus maximus m.

Figure 19.8 A ventral view of the neck and thorax.

1. Digastric m.
2. Mylohyoid m.
3. Sternomastoid m.
4. Clavotrapezius m.
5. Masseter m.
6. Clavobrachialis m.
7. Pectoantebrachialis m.
8. Pectoralis major m.

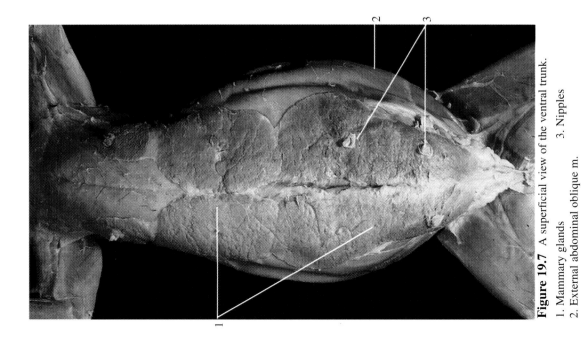

Figure 19.7 A superficial view of the ventral trunk.

1. Mammary glands
2. External abdominal oblique m.
3. Nipples

Figure 19.10 An anterior view of the brachium and antebrachium.

1. Masseter m.
2. Sternomastoid m.
3. Clavobrachialis m.
4. Pectoantebrachialis m.
5. Pectoralis major m.
6. Pectoralis minor m.
7. Extensor carpi radialis longus m.
8. Flexor carpi ulnaris m.
9. Pronator teres m.
10. Epitrochlearis

Figure 19.9 A lateral view of the shoulder and brachium.

1. Acromiotrapezius m.
2. Levator scapulae ventralis m.
3. Spinodeltoid m.
4. Latissimus dorsi m.
5. Long head of triceps brachii m.
6. Clavobrachialis m.
7. Lateral head of triceps brachii m.
8. Clavotrapezius m.
9. Auricular m.
10. Acromiodeltoid m.

Figure 19.12 An anterior view of the trunk.

1. Sternomastoid m.
2. Scalenus anterior m.
3. Scalenus posterior m.
4. Epitrochlearis m.
5. Transverse costarum m.
6. Pectoralis minor m. (cut)
7. Rectus abdominis m.
8. Xiphihumeralis m. (cut)
9. Pectoralis minor (cut)
10. Epitrochlearis m.
11. Subscapularis m.
12. Scalenus medius m.
13. Serratus anterior m.
14. Latissimus dorsi m. (cut)
15. External abdominal oblique m.

Figure 19.11 A posterior view of the brachium and antebrachium.

1. Brachioradials m.
2. Extensor digitorum lateralis m.
3. Extensor digitorum communis m.
4. Extensor carpi radialis longus m.
5. Lateral head of triceps brachii m.
6. Long head of triceps brachii m.
7. Spinodeltoid m.
8. Latissimus dorsi m.
9. Clavobrachialis m.
10. Acromiotrapezius m.

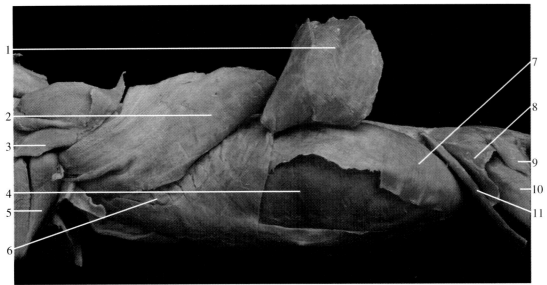

Figure 19.13 A lateral view of the trunk.

1. External abdominal oblique m.
2. Latissimus dorsi m.
3. Spinodeltoid m.
4. Transverse abdominis m.
5. Long head of triceps brachii m.
6. Serratus anterior m.

7. Internal abdominal oblique m.
8. Tensor fascia latae
9. Caudofemoris m.
10. Vastus lateralis m.
11. Sartorius m.

Figure 19.14 A lateral view of the superficial thigh.

1. Internal abdominal oblique m.
2. External abdominal oblique m.
3. Tensor fascia latae (cut)
4. Vastus lateralis m.
5. Biceps femoris m.

6. Sartorius m.
7. Gluteus medius m.
8. Gluteus maximus m.
9. Caudofemoris m.
10. Caudal m.
11. Semitendinosus m.

Figure 19.15 A medial view of the thigh and leg.

1. Rectus abdominus m.
2. Adductor longus m.
3. Adductor femoris m.
4. Semimembranosus m.
5. Gracilis m. (cut)
6. Tendo calcaneus (Achilles tendon)
7. Sartorius m.

8. Tensor fascia latae m.
9. Vastus lateralis m.
10. Rectus femoris m.
11. Vastus medialis m.
12. Flexor digitorum longus m.
13. Tibialis anterior m.

Figure 19.17 Intact viscera.

1. Right brachiocephalic vein
2. Superior vena cava
3. Right lung
4. Greater omentum
5. Left brachiocephalic vein
6. Left lung
7. Heart
8. Liver
9. Stomach
10. Mesentery
11. Small intestine

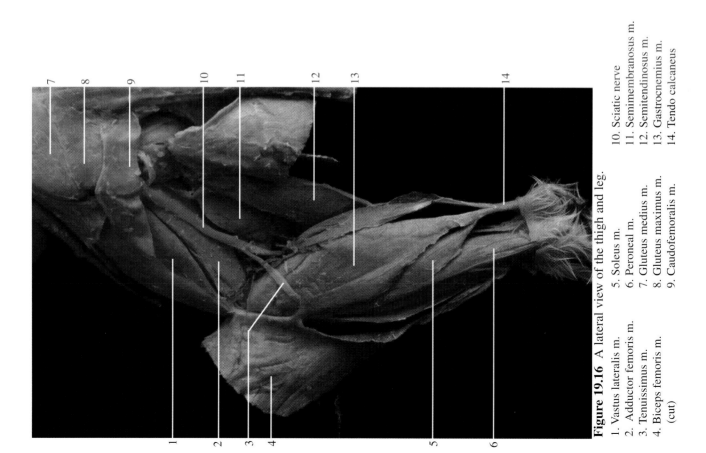

Figure 19.16 A lateral view of the thigh and leg.

1. Vastus lateralis m.
2. Adductor femoris m.
3. Tenuissimus m.
4. Biceps femoris m. (cut)
5. Soleus m.
6. Peroneal m.
7. Gluteus medius m.
8. Gluteus maximus m.
9. Caudofemoralis m.
10. Sciatic nerve
11. Semimembranosus m.
12. Semitendinosus m.
13. Gastrocnemius m.
14. Tendo calcaneus

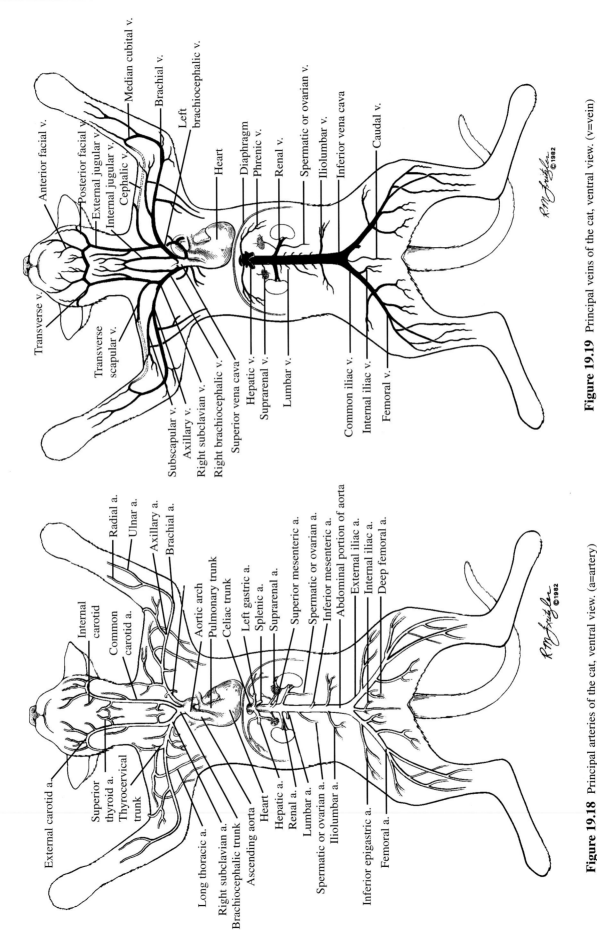

Figure 19.19 Principal veins of the cat, ventral view. (v=vein)

Figure 19.18 Principal arteries of the cat, ventral view. (a=artery)

Figure 19.22 Cat heart within pericardium.

1. Right brachiocephalic vein
2. Superior vena cava
3. Aortic arch
4. Heart
5. Liver
6. Left lobe of lung

Figure 19.20 The cat heart and associated vessels.

Right common carotid artery
Left common carotid artery
Left subclavian artery
Aortic arch
Pulmonary trunk
Pulmonary arteries
Left atrium
Descending aorta
Left ventricle
Brachiocephalic vein
Right subclavian artery
Brachiocephalic trunk
Superior vena cava
Azygos vein
Right atrium
Right ventricle

©1982

Figure 19.21 Internal anatomy of the cat heart.

Brachiocephalic trunk
Aortic arch
Pulmonary trunk
Pulmonary veins
Pulmonary valve
Left (atrioventricular) bicuspid valve
Left atrium
Left ventricle
Apex of heart
Superior vena cava
Right atrium
Right (atrioventricular) tricuspid valve
Chordae tendineae
Papillary muscle
Right ventricle
Interventricular septum

Figure 19.24 Heart and surrounding structures.

1. Trachea
2. Common carotid arteries
3. Axillary vein
4. Heart (cut)
5. Left ventricle
6. Vagus nerve
7. External jugular vein
8. Left subclavian vein
9. Superior vena cava
10. Brachiocephalic trunk
11. Thoracic aorta

Figure 19.23 An internal view of the heart.

1. Trachea
2. Right common carotid artery
3. Right subclavian vein
4. Right brachiocephalic vein
5. Superior vena cava
6. Right atrium
7. Right lung
8. Right ventricle
9. Left common carotid artery
10. Internal jugular vein
11. Brachiocephalic artery
12. Right common carotid artery
13. Aortic arch
14. Left lung
15. Left ventricle

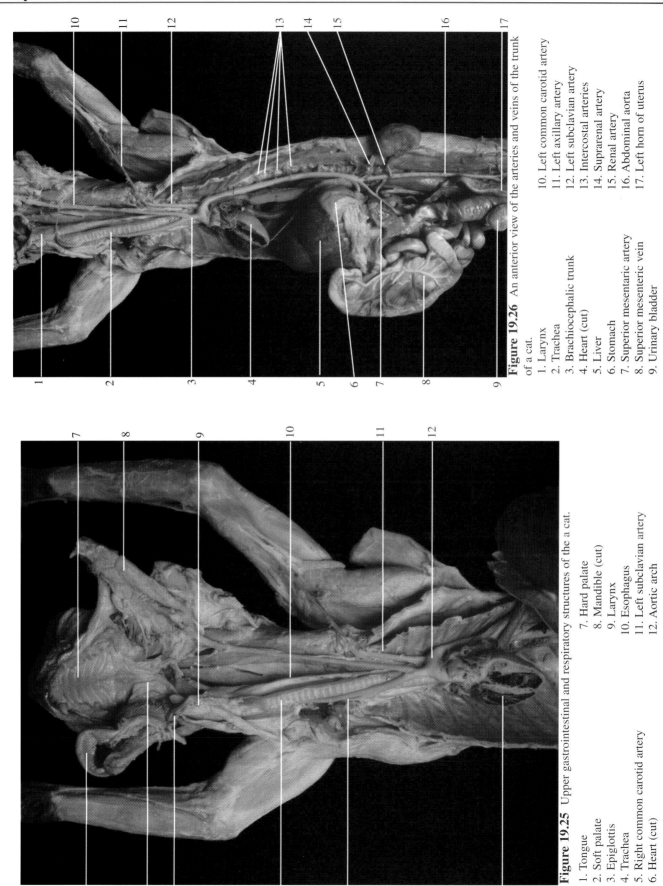

Figure 19.26 An anterior view of the arteries and veins of the trunk of a cat.

1. Larynx
2. Trachea
3. Brachiocephalic trunk
4. Heart (cut)
5. Liver
6. Stomach
7. Superior mesenteric artery
8. Superior mesenteric vein
9. Urinary bladder
10. Left common carotid artery
11. Left axillary artery
12. Left subclavian artery
13. Intercostal arteries
14. Suprarenal artery
15. Renal artery
16. Abdominal aorta
17. Left horn of uterus

Figure 19.25 Upper gastrointestinal and respiratory structures of the a cat.

1. Tongue
2. Soft palate
3. Epiglottis
4. Trachea
5. Right common carotid artery
6. Heart (cut)
7. Hard palate
8. Mandible (cut)
9. Larynx
10. Esophagus
11. Left subclavian artery
12. Aortic arch

Figure 19.26 An anterior view of the viscera.

1. Heart
2. Esophagus
3. Descending aorta
4. Liver
5. Greater omentum
6. Diaphragm
7. Small intestine
8. Mesentery

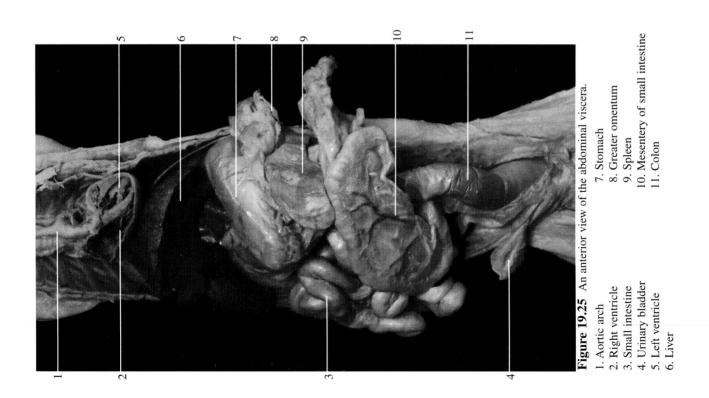

Figure 19.25 An anterior view of the abdominal viscera.

1. Aortic arch
2. Right ventricle
3. Small intestine
4. Urinary bladder
5. Left ventricle
6. Liver
7. Stomach
8. Greater omentum
9. Spleen
10. Mesentery of small intestine
11. Colon

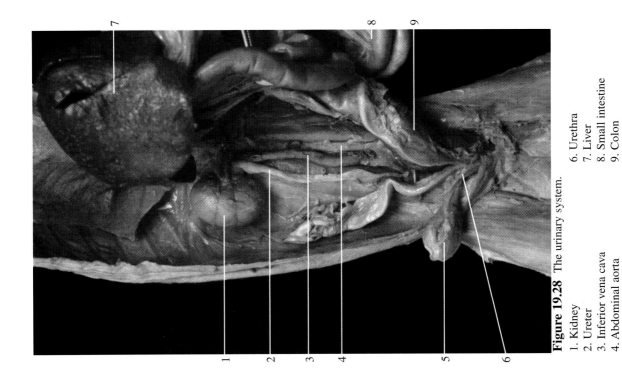

Figure 19.28 The urinary system.

1. Kidney
2. Ureter
3. Inferior vena cava
4. Abdominal aorta
5. Urinary bladder
6. Urethra
7. Liver
8. Small intestine
9. Colon

Figure 19.27 An anterior view of the deep structures of the trunk.

1. Right common carotid artery
2. Vagus nerve
3. Thoracic aorta
4. Intercostal artery
5. Kidney
6. Ureter
7. Left brachiocephalic vein
8. Superior vena cava
9. Heart
10. Celiac trunk
11. Superior mesenteric artery
12. Liver
13. Spleen
14. Small intestine

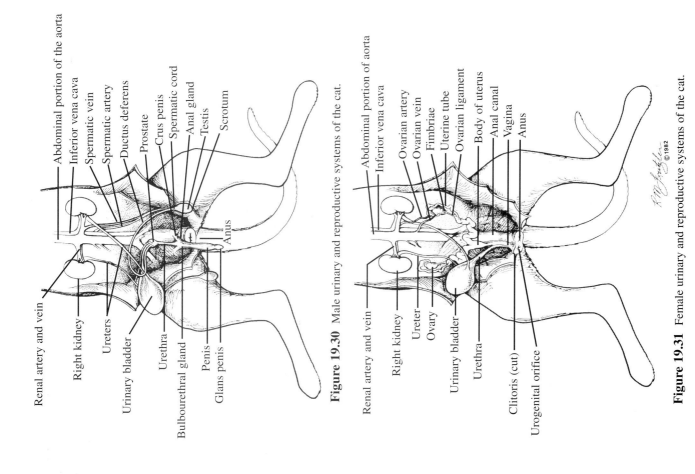

Abdominal portion of the aorta
Inferior vena cava
Spermatic vein
Spermatic artery
Ductus deferens
Prostate
Crus penis
Spermatic cord
Anal gland
Testis
Scrotum

Renal artery and vein
Right kidney
Ureters
Urinary bladder
Urethra
Bulbourethral gland
Penis
Glans penis
Anus

Figure 19.30 Male urinary and reproductive systems of the cat.

Abdominal portion of aorta
Inferior vena cava
Ovarian artery
Ovarian vein
Fimbriae
Uterine tube
Ovarian ligament
Body of uterus
Anal canal
Vagina
Anus

Renal artery and vein
Right kidney
Ureter
Ovary
Urinary bladder
Urethra
Clitoris (cut)
Urogenital orifice

Figure 19.31 Female urinary and reproductive systems of the cat.

Figure 19.29 The urinary and female reproductive structures.

1. Renal cortex
2. Renal medulla
3. Renal pelvis
4. Ureter
5. Horn of uterus
6. Urinary bladder

7. Liver
8. Small intestine
9. Inferior vena cava
10. Left horn of uterus
11. Colon
12. Urethra

Figure 19.32 Abdominal cavity of a pregnant cat.
1. Greater omentum
2. Right horn of uterus

Figure 19.33 Abdominal cavity of a pregnant cat.
1. Uterine wall (cut)
2. Amniotic sac enclosing fetus
3. Fetus

Pig Dissection

20

Because much can be learned from dissecting embalmed fetal pig specimens, they are frequently utilized in anatomy laboratories. Fetal pigs are purchased from biological supply houses and are specially prepared for dissection. Excess embalming fluid should be drained from the packaged specimen prior to dissection.

Examine your specimen and identify the *umbilical cord* attached to the ventral surface of the abdomen. Locate the two rows of *teats* that extend the length of the abdomen. Determine the sex of your specimen. A male has a *scrotal sac* in the pelvic region of the body between the hind legs and a *urogenital opening* just caudal to the umbilical cord. The *penis* can be palpated as a muscular tubular structure just underneath the skin along the midline proceeding caudally from the urogenital opening. A female has a small fleshy *genital papilla* projecting from the urogenital opening, which is located immediately ventral to the *anal opening*.

Before the muscles and viscera of a fetal pig can be studied, the specimen's skin has to be removed according to the following suggested guidelines.

1. Place your specimen on a dissecting tray ventral side up. Using a sharp scalpel, make a shallow incision through the skin extending from the chin caudally to the umbilical cord. Carefully continue your cut around one side of the umbilical cord. If your specimen is a male, make a diagonal cut from the umbilical cord to the scrotum. If a female, continue a midventral incision from the umbilical cord to the genital papilla. Make an incision around the genitalia and tail.

2. From the midventral incision, extend an incision down the medial surfaces of the forelegs to the hoofs and then do the same for the skin of the hindlegs. Make circular incisions around each of the hoofs. Following the ventral borders of the lower jaws, make extended cuts from the chin dorsolaterally to just below the ears.

3. Grasp the cut edge of the skin and carefully remove it from your specimen. If the skin is difficult to remove, grasp the cut edge of the skin with one hand and push on the muscle with the thumb of the other hand.

4. After the specimen is skinned, the muscles can be seen more easily if the moisture on them is sponged away with a paper towel. The muscles of a fetal pig are extremely delicate and as you proceed to dissect your specimen, make certain that you separate the muscles along their natural boundaries. When transection of a muscle is necessary, carefully isolate the muscle from its attached connective tissue and make a clean cut across the belly of the muscle, leaving the origin and insertion intact.

5. At the end of the laboratory period, wrap your specimen in muslin cloth and store it in a tight, heavy-duty plastic bag. Discard the skin that was removed from your specimen, and the plastic shipment bag. Wet your specimen from time to time with a preservative solution (usually 2-3% phenol). Caution is necessary when using a phenol wetting solution as it is caustic and poisonous if misused or used in a concentrated form.

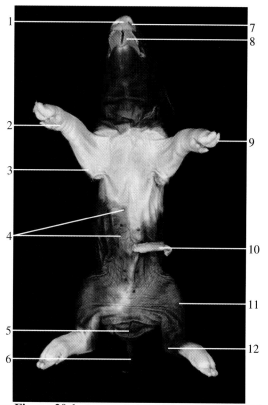

Figure 20.1 A ventral view of the surface anatomy of the fetal pig.

1. Nose	5. Scrotum	9. Digit
2. Wrist	6. Tail	10. Umbilical cord
3. Elbow	7. Nostril	11. Knee
4. Teats	8. Tongue	12. Ankle

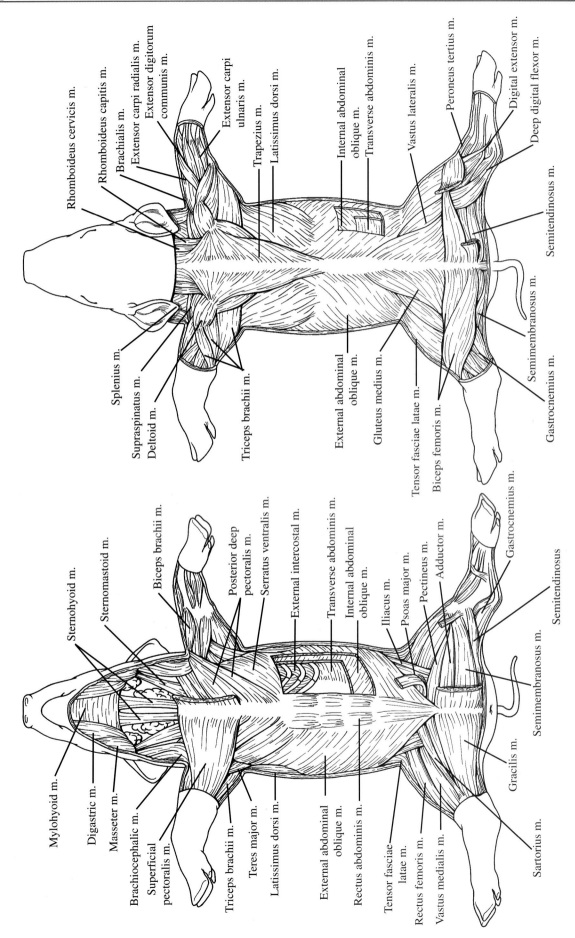

Rhomboideus cervicis m.
Rhomboideus capitis m.
Brachialis m.
Extensor carpi radialis m.
Extensor digitorum communis m.
Extensor carpi ulnaris m.
Trapezius m.
Latissimus dorsi m.
Internal abdominal oblique m.
Transverse abdominis m.
Vastus lateralis m.
Peroneus tertius m.
Digital extensor m.
Deep digital flexor m.
Semitendinosus m.
Semimembranosus m.
Gastrocnemius m.
Biceps femoris m.
Tensor fasciae latae m.
Gluteus medius m.
External abdominal oblique m.
Triceps brachii m.
Supraspinatus m.
Deltoid m.
Splenius m.

Figure 20.3 A dorsal view of the muscles of the fetal pig.

Sternohyoid m.
Sternomastoid m.
Biceps brachii m.
Posterior deep pectoralis m.
Serratus ventralis m.
External intercostal m.
Transverse abdominis m.
Internal abdominal oblique m.
Iliacus m.
Psoas major m.
Pectineus m.
Adductor m.
Gastrocnemius m.
Semitendinosus
Semimembranosus m.
Gracilis m.
Vastus medialis m.
Rectus femoris m.
Tensor fasciae latae m.
Rectus abdominis m.
External abdominal oblique m.
Latissimus dorsi m.
Teres major m.
Triceps brachii m.
Superficial pectoralis m.
Brachiocephalic m.
Masseter m.
Digastric m.
Mylohyoid m.
Sartorius m.

Figure 20.2 A ventral view of the muscles of the fetal pig.

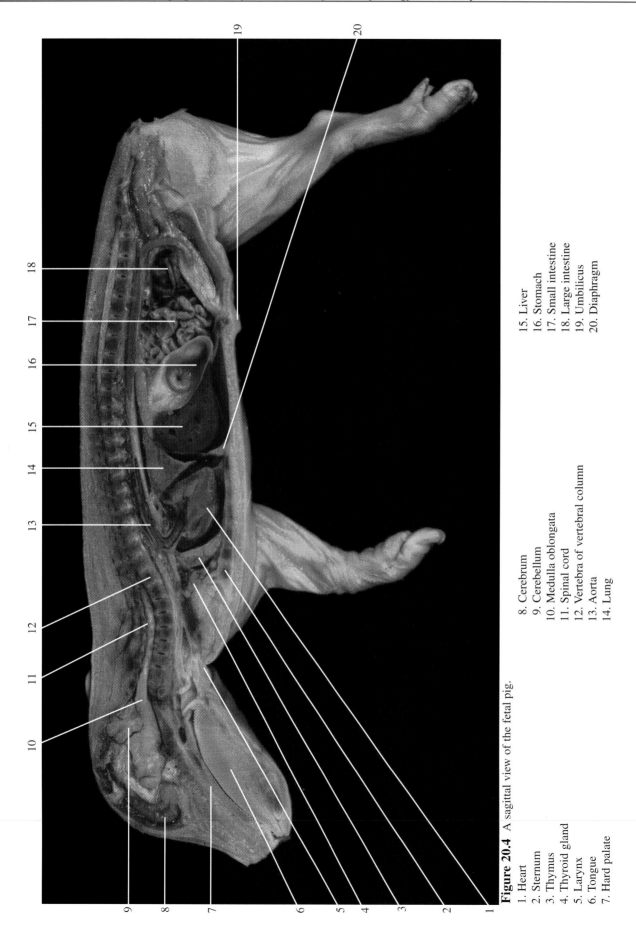

Figure 20.4 A sagittal view of the fetal pig.

1. Heart
2. Sternum
3. Thymus
4. Thyroid gland
5. Larynx
6. Tongue
7. Hard palate

8. Cerebrum
9. Cerebellum
10. Medulla oblongata
11. Spinal cord
12. Vertebra of vertebral column
13. Aorta
14. Lung

15. Liver
16. Stomach
17. Small intestine
18. Large intestine
19. Umbilicus
20. Diaphragm

Figure 20.6 Deep viscera and associated structures.

1. Liver
2. Small intestine
3. Larynx
4. Heart
5. Lung
6. Kidney

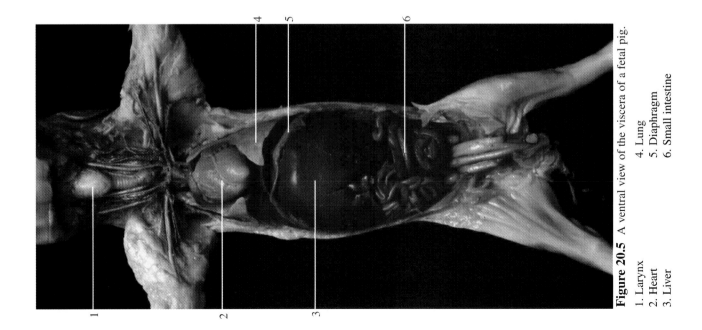

Figure 20.5 A ventral view of the viscera of a fetal pig.

1. Larynx
2. Heart
3. Liver
4. Lung
5. Diaphragm
6. Small intestine

Figure 20.7 Thorax and neck regions of the fetal pig.

1. Larynx   5. Thymus
2. Thyroid gland   6. Heart
3. Lung   7. Lung
4. Liver

Figure 20.8 A ventral view of the abdominal cavity of a fetal pig.

1. Diaphragm   5. Undescended testis
2. Liver   6. Umbilical artery
3. Umbilical vein   7. Urinary bladder
4. Small intestine

Figure 20.9 Abdominal organs of the fetal pig.

1. Liver
2. Small intestine
3. Kidney
4. Ureter
5. Ductus (vas) deferens

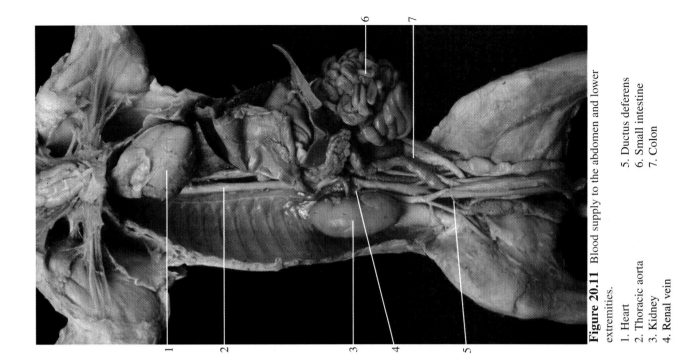

Figure 20.11 Blood supply to the abdomen and lower extremities.

1. Heart
2. Thoracic aorta
3. Kidney
4. Renal vein
5. Ductus deferens
6. Small intestine
7. Colon

Figure 20.10 Arteries and veins of the neck and thoracic region.

1. Larynx
2. Right auricle
3. Right lung
4. Trachea
5. Common carotid artery
6. Left auricle
7. Heart
8. Left lung
9. Diaphragm

Figure 20.12 Urogenital system of the fetal pig.

1. Kidney
2. Inferior vena cava
3. Ureter
4. Rectum (cut)
5. Partially dissected testis
6. Renal vein
7. Descending aorta
8. Ductus deferens
9. Urinary bladder
10. Umbilical vein
11. Epididymis

Figure 20.13 Urogenital system of the fetal pig.

1. Umbilical cord
2. Right kidney
3. Ureter
4. Umbilical artery
5. Urinary bladder
6. Penis
7. Spermatic cord
8. Right testis
9. Epididymis

Figure 20.14 General structures of the fetal pig brain. Because the cerebrum is less defined in pigs, the regions are not known as lobes as they are in humans.

1. Occipital region of cerebrum
2. Cerebellum
3. Medulla oblongata
4. Spinal cord
5. External acoustic meatus
6. Longitudinal fissure
7. Parietal region of cerebrum
8. Frontal region of cerebrum
9. Temporal region of cerebrum
10. Eye

Rat Dissection

Figure A.1 Ventral view of the rat viscera.

1. Trachea	4. Submandibular gland
2. Liver	5. Heart
3. Small intestine	6. Diaphragm

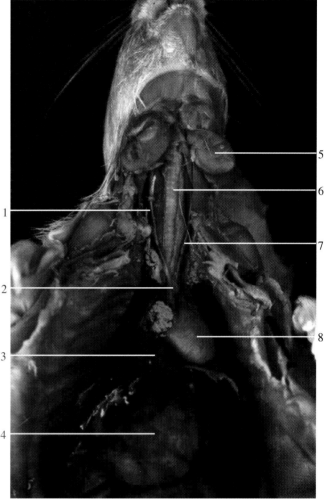

Figure A.2 Neck region of the rat.

1. Common carotid artery	5. Submandibular gland
2. Aortic arch	6. Trachea
3. Lung	7. Common carotid artery
4. Liver	8. Heart

Figure A.3 Structures of the rat abdomen.
1. Small intestine
2. Caudal vena cava
3. Colon
4. Horn of uterus
5. Liver
6. Kidney
7. Dorsal aorta
8. Urinary bladder

Figure A.4 Trunk of the rat.
1. Small intestine
2. Caudal vena cava
3. Horn of uterus
4. Liver
5. Kidney
6. Abdominal aorta
7. Colon
8. Urinary bladder

Glossary of Prefixes and Suffixes

Element	Definition and Example	Element	Definition and Example	Element	Definition and Example
a-	absent, deficient or without: atrophy	chrom-	color: chromocyte	fiss-	split: fissure
ab-	off, away from: abduct	-cid	destroy: germicide	for-	opening: foramen
abdomin-	abdomen	circum-	around: circumduct	-form	shape: fusiform
-able	capable of: viable	-cis	cut, kill: excision		
ac-	toward, to: actin	co-	together: copulation	gastro-	stomach: gastrointestinal
acou-	hear, acoustic	coel-	hollow cavity: coelom	-gen	an agent that produces or originates: pathogen
ad-	denoting to, toward: adduct	con-	with, together: congenital		
af-	movement toward a central point: afferent artery	contra-	against, opposite: contraception	-genic	produced from, producing: carcinogenic
alba-	pale or white: linea alba	corn-	denoting hardness: cornified	gloss-	tongue: glossopharyngeal
-alg	pain: neuralgia	corp-	body: corpus	glyco-	sugar: glycosuria
ambi-	both: ambidextrous	crypt-	hidden: cryptorchism	-gram	a record, recording: myogram
angi-	pertaining to vessel: angiology	cyan-	blue color: cyanosis	gran-	grain, particle: agranulocyte
ante-	before: antebrachium	cysti-	sac or bladder: cystoscope	-graph	instrument for recording: electrocardiograph
anti-	against: anticoagulant	cyto-	cell: cytology		
aqua-	water: aqueous			grav-	heavy: gravid
archi-	to be first: archeteron	de-	down, from: descent	gyn-	female sex: gynocology
arthri-	joint: arthritis	derm-	skin: dermatology		
-asis	condition or state of: homeostasis	di-	two: diarthrotic	hema(o)-	blood: hematology
aud-	pertaining to ear: auditory	dipl-	double: diploid	haplo-	simple or single: haploid
auto-	self: autolysis	dis-	apart, away from: disarticulate	hemi-	half: hemiplegia
		duct-	lead, conduct: ductus deferens	hepat-	liver: hepatic portal
bi-	two: bipedal	dur-	hard: dura mater	hetero-	other, different: heterosexual
bio-	life: biology	-dynia	pain	histo-	webb, tissue: histology
blast-	generative or germ bud: osteoblast	dys-	bad, difficult, painful: dysentery	holo-	whole, entire: holocrine
brachi-	arm: brachialis			homo-	same, alike: homologous
brachy-	short: brachydont	e-	out, from: eccrine	hydro-	water: hydrocoel
brady-	slow: bradycardia	ecto-	outside, outer, external: ectoderm	hyper-	beyond, above, excessive: hypertension
bucc-	cheek: buccal cavity	-ectomy	surgical removal: tonsillectomy	hypo-	under, below: hypoglycemia
cac-	bad, ill: cachexia	ede-	swelling: edema	-ia	state or condition: hypoglycemia
calci-	stone: calculus	-emia	pertaining to a condition of the blood: lipemia	-iatrics	medical specialties: pediatrics
capit-	head: capitis	end-	within: endoderm	idio-	self, separate, distinct: idiopathic
carcin-	cancer: carcinogenic	entero-	intestine: enteritis		
cardi-	heart: cardiac	epi-	upon, in addition: epidermis	ilio-	ilium: iliosacral
caud-	tail: cauda equina	erythro-	red: erythrocyte	infra-	beneath: infraspinatus
cata-	lower, under, against: catabolism	ex-	out of: excise	inter-	among, between
-coel	swelling, and enlarged space or cavity: blastocoele	exo-	outside: exocrine	intra-	inside, within: intracellular
		extra-	outside of, beyond, in addition: extracellular	-ion	process: acromion
cephal-	head: cephalis			iso-	equal, like: isotonic
cerebro-	brain: cerebrospinal fluid	fasci-	band: fascia	-ism	condition or state: rheumatism
chol-	bile: cholic	febr-	fever: febrile	-itis	inflammation: meningitis
chondr-	cartilage: chondrocyte	-ferent	bear, carry: efferent arteriole	labi-	lip: labium majus
				lacri-	tears: nasolacrimal
				later-	side: lateral

Glossary of Prefixes and Suffixes

Element	Definition and Example	Element	Definition and Example	Element	Definition and Example
leuc-	white: leucocyte	phag-	to eat: phagocyte	tens-	stretch: tensor fascia lata
lip-	fat: lipid	-phil	have an affinity for: neutrophil	tetra-	four: tetrad
-logy	science of: morphology	phlebo-	vein: phlebitis	therm-	heat: thermogram
-lysis	solution, dissolve: hemolysis	-phobe	abnormal fear, dread: hydrophobia	thorac-	chest: thoracic cavity
				thrombo-	lump, clot: thrombocyte
macro-	large, great: macrophage	-plasty	reconstruction of: rhinoplasty	-tomy	cut: appendectomy
mal-	bad, abnormal, disorder: malignant	platy-	flat, side: platysma	tox-	poison: toxic
medi-	middle: medial	-plegia	stroke, paralysis: paraplegia	tract-	draw, drag: traction
mega-	great, large: mega karyocyte	-pnea	to breathe: apnea	trans-	across, over: transfuse
meso-	middle or moderate: mesoderm	pneumato-	breathing: pneumonia	tri-	three: trigone
meta-	after, beyond: metatarsal	pod-	foot: podiatry	trich-	hair: trichology
micro-	small: microtome	-poieis	formation of: hematopoiesis	-trophy	a state relating to nutrition: hypertrophy
mito-	thread: mitosis	poly-	many, much: polyploid		
mono-	alone, one, single: monocyte	post-	after, behind: post natal	-tropic	turning toward, changing: gonadotropic
mons-	mountain: mons pubis	pre-	before in time or place: prenatal		
morph-	form, shape: morphology	prim-	first: primitive	ultra-	beyond, excess: ultrasonic
multi-	many, much: multinuclear	pro-	before in time or place: prosect	uni-	one: unicellular
myo-	muscle: myology	proct-	anus: proctology	uro-	urine, urinary organs or tract: uroscope
		pseudo-	false: pseudostratified		
narc-	numbness, stupor: narcotic	psycho-	mental: psychology	-uria	urine: polyuria
neo-	new, young: neonatal	pyo-	pus: pyoculture		
necro-	corpse, dead: necrosis			vas-	vessel: vasoconstriction
nephro-	kidney: nephritis	quad-	fourfold: quadriceps femoris	vermi-	worm: vermiform
neuro-	nerve: neurolemma			viscer-	organ: visceral
noto-	back: notochord	re-	back, again: repolarization	vit-	life: vitamin
		rect-	straight: rectus abdominis		
ob-	against, toward, in front of: obturator	reno-	kidney: renal	zoo-	animal: zoology
oc-	against: occlusion	rete-	network: retina	zygo-	union, join: zygote
-oid	resembling, likeness: sigmoid	retro-	backward: retroperitoneal		
oligo-	few, small: oligodendrocyte	rhin-	nose: rhinitis		
-oma	tumor: lymphoma	-rrhage	excessive flow: hemorrhage		
oo-	egg: oocyte	-rrhea	flow, or discharge: diarrhea		
or-	mouth: oral				
orchi-	testicles: cryptorchidism	sanguin-	blood: sanguiferous		
-ory	pertaining to: sensory	sarc-	flesh: sarcoplasm		
osteo-	bone: osteoblast	-scope	instrument for examination of a part: stethoscope		
-ose	full of: adipose	-sect	cut: dissect		
oto	ear: otolith	semi-	half: semilunar		
ovo-	egg: ovum	serrate-	saw-edged: serratus anterior		
		-sis	state or condition: dialysis		
para-	give birth to, bear: parturition	steno-	narrow: stenohaline		
para-	near, beyond, beside: paranasal	-stomy	surgical opening: tracheotomy		
path-	disease, that which undergoes sickness: pathology	sub-	under, beneath, below: subcutaneous		
-pathy	abnormality, disease: neuropathy	super-	above, beyond, upper: superficial		
ped-	children: pediatrician	supra-	above, over: suprarenal		
pen-	need, lack: penicillin	syn (sym)	together, joined, with: synapse		
-penia	deficiency: thrombocytopenia				
per-	through: percutaneous	tachy-	swift, rapid: tachometer		
peri-	near, around: pericardium	tele-	far: telencephalon		

Glossary

abdomen (ab-do´men): the portion of the trunk located between the diaphragm and the pelvis; contains the abdominal cavity and its visceral organs.

abduction (ab-duk´shun): a movement away from the axis or midline of the body; opposite of adduction; a movement of a digit away from the axis of a limb.

acapnia (ah-kap´ne-ah): a decrease in normal amount of CO_2 in the blood.

accommodation (ah-kom-o-da´shun): a change in the shape of the lens of the eye so that vision is more acute; the focusing for various distances.

acetone (as´e-tone): an organic compound that may be present in the urine of diabetics; also called ketone bodies.

Achilles tendon (ah-kil´ez): see *tendo calcaneus*.

acidosis (as-i-do´sis): a disorder of body chemistry in which the alkaline substances of the blood are reduced below normal.

actin (ak´tin): a protein in muscle fibers that together with myosin is responsible for contraction.

acoustic (ah-koos´tik): referring to sound or the sense of hearing.

adduction (ah-duk´shun): a movement toward the axis or midline of the body; opposite of abduction; a movement of a digit toward the axis of a limb.

adenohypophysis (ad´e-no-hi-pof´i-sis): anterior pituitary gland.

adenoid (ad´e-noid): paired lymphoid structures in the nasopharynx; also called pharyngeal tonsils.

adenosine triphosphate (ATP) (ah-den´o-sen tri-fos´fate): a chemical compound that provides energy for cellular use.

adipose (ad´e-pose): fat, or fat-containing, such as adipose tissue.

adrenal glands (ah-dre´nal): endocrine glands; one superior to each kidney; also called *suprarenal glands*.

aerobic (a-er-o´bik): requiring free O_2 for growth and metabolism as in the case of certain bacteria called *aerobes*.

allantois (ah-lan´to-is): an extraembryonic membranous sac that forms blood cells and gives rise to the fetal umbilical arteries and vein. It also contributes to the formation of the urinary bladder.

alveolus (al-ve´o-lus): an individual aircapsule within the lung. Alveoli are the basic functional units of respiration. Also, the socket that secures a tooth.

amnion (am´ne-on): a membrane that surrounds the fetus to contain the amniotic fluid.

amphiarthrosis (am´fe-ar-thro´sis): a slightly moveable joint in a functional classification of joints.

anatomical position (an´ah-tom´e-kal): an erect body stance with the eyes directed forward, the arms at the sides, and the palms of the hands facing forward.

anatomy (ah-nat´o-me): the branch of science concerned with the structure of the body and the relationship of its organs.

antebrachium (an´te-bra´ke-um): the forearm.

anterior (ventral) (an-te´re-or): toward the front; the opposite of *posterior (dorsal)*.

antigen (an´te-jen): a substance which causes cells to produce antibodies.

anus (a´nus): the terminal end of the GI tract, opening of the anal canal.

aorta (a-or´tah): the major systemic vessel of the arterial portion of the circulatory system, emerging from the left ventricle.

apocrine gland (ap´o-krin): a type of sweat gland that functions in evaporative cooling.

appendix (ah-pen´diks): a short pouch that attaches to the cecum.

aqueous humor (a´kwe-us hu´mor): the watery fluid that fills the anterior and posterior chambers of the eye.

arachnoid mater (ah-rak´noid): the weblike middle covering (meninx) of the central nervous system.

arbor vitae (ar´bor vi´tah): the branching arrangement of white matter within the cerebellum.

areola (ah-re´o-lah): the pigmented ring around the nipple.

artery (ar´ter-e): a blood vessel that carries blood away from the heart.

articular cartilage (ar-tik´u-lar ker´ti-lij): a hyaline cartilaginous covering over the articulating surface of bones of synovial joints.

ascending colon (ko´lon): the portion of the large intestine between the cecum and the right colic (hepatic) flexure.

atom (at´om): the smallest unit of an element that can exist and still have the properties of the element; collectively, atoms form molecules in a compound.

atrium (a´tre-um): either of two superior chambers of the heart that receive venous blood.

atrophy (at´ro-fe): a wasting away or decrease in size of a cell or organ.

auditory tube (aw´di-to´re): a narrow canal that connects the middle ear chamber to the pharynx; also called the *eustachian canal*.

autonomic (aw-to-nom´ik): self-governing; pertaining to the division of the nervous system which controls involuntary activities.

axilla (ak-sil´ah): the depressed hollow under the arm; the armpit.

axon (ak´son): The elongated process of a neuron (nerve cell) that transmits an impulse away from the cell body.

basement membrane: a thin sheet of extracellular substance to which the basal surfaces of membranous epithelial cells are attached.

basophil (ba´so-fil): a granular leukocyte that readily stains with basophilic dye.

belly: the thickest circumference of a skeletal muscle.

benign: (be-nine´): nonmalignant; a confined tumor.

blastula (blas´tu-lah): an early stage of prenatal development between the morula and embryonic stages.

blood: the fluid connective tissue that circulates through the cardiovascular system to transport substances throughout the body.

bolus (bo´lus): a moistened mass of food that is swallowed from the oral cavity into the pharynx.

bone: an organ composed of solid, rigid connective tissue, forming a component of the skeletal system.

Bowman's capsule (bo´manz kap´sul): see *glomerular capsule*.

brain: the enlarged superior portion of the central nervous system, located in the cranial cavity of the skull.

brain stem: the portion of the brain consisting of the medulla oblongata, pons, and midbrain.

bronchial tree (brong´ke-al): the bronchi and their branching bronchioles.

bronchiole (brong´ke-ol): a small division of a bronchus within the lung.

bronchus (bron´kus): a branch of the trachea that leads to a lung.

buccal cavity (buk´al): the mouth, or oral cavity.

bursa (ber´sah): a saclike structure filled with synovial fluid, which occurs around joints.

buttock (but´ok): the rump or fleshy mass on the posterior aspect of the lower trunk, formed primarily by the gluteal muscles.

calorie (kal´o-re): the unit of heat required to raise the temperature of one gram of water one degree centigrade.

calyx (ka´liks): a cup-shaped portion of the renal pelvis that encircles a renal papilla.

cancellous bone (kan´se-lus): spongy bone; bone tissue with a latticelike structure.

capillary (kap´i-lar´e): a microscopic blood vessel that connects an arteriole and a venule; the functional unit of the circulatory system.

carcinogenic (kar-si-no-jen´ik): stimulating or causing the growth of a malignant tumor, or cancer.

carpus (kar´pus): the proximal portion of the hand that contains the eight carpal bones.

cartilage (kar´ti-lij): a type of connective tissue with a solid elastic matrix.

caudal (kaw´dal): referring to a position more toward the tail.

cecum (se´kum): the pouchlike portion of the large intestine to which the ileum of the small intestine is attached.

cell: the structural and functional unit of an organism; the smallest structure capable of performing all the functions necessary for life.

central nervous system (CNS): the brain and the spinal cord.

centrosome (sen´tro-som): a dense body near the nucleus of a cell that contains a pair of centrioles.

cerebellum (ser´e-bel´um): the portion of the brain concerned with the coordination of movements and equilibrium.

cerebrospinal fluid (ser´e-bro-spi´nal): a fluid that buoys and cushions the central nervous system.

cerebrum (ser´e-brum): the largest portion of the brain, composed of the right and left hemispheres.

cervical (ser´vi-kal): pertaining to the neck or a necklike portion of an organ.

choanae (ko-a´na): the two posterior openings from the nasal cavity into the nasopharynx.

cholesterol (ko-les´ter-ol): an organic fat-like compound found in animal fat, bile, blood, liver, and other parts of the body.

chondrocyte (kon´dro-site): a cartilage cell.

chorion (ko´re-on): An extraembryonic membrane that participates in the formation of the placenta.

choroid (ko´roid): the vascular, pigmented middle layer of the wall of the eye.

chromosome (kro´mo-som): structure in the nucleus that contains the genes for genetic expression.

chyme (kime): The mass of partially digested food that passes from the stomach into the duodenum of the small intestine.

cilia (sil´e-ah): microscopic, hairlike processes that move in a wavelike manner on the exposed surfaces of certain epithelial cells.

ciliary body (sil´e-er´e): a portion of the choroid layer of the eye that secretes aqueous humor and contains the ciliary muscle.

circumduction (ser´kum-duk´shun): a conelike movement of a body part, such that the distal end moves in a circle while the proximal portion remains relatively stable.

clitoris (kli´to-ris): a small, erectile structure in the vulva of the female.

cochlea (kok´le-ah): the spiral portion of the inner ear that contains the spiral organ (organ of Corti).

ceolom (se´lom): the abdominal cavity.

colon (ko´lon): the first portion of the large intestine.

common bile duct: a tube that is formed by the union of the hepatic duct and cystic duct, transports bile to the duodenum.

compact bone: tightly packed bone that is superficial to spongy bone; also called *dense bone*.

condyle (kon´dile): a rounded process at the end of a long bone that forms an articulation.

connective tissue: one of the four basic tissue types within the body. It is a binding and supportive tissue with abundant matrix.

cornea (kor´ne-ah): the transparent convex, anterior portion of the outer layer of the eye.

cortex (kor´teks): the outer layer of an organ such as the convoluted cerebrum, adrenal gland, or kidney.

costal cartilage (kos´tal): the cartilage that connects the ribs to the sternum.

cranial (kra´ne-al): pertaining to the cranium.

cranial nerve: one of twelve pairs of nerves that arise from the inferior surface of the brain.

cranium (kra´ne-um): the bones of the skull that enclose the brain and support the organs of sight, hearing, and balance.

cytology (si-tol´o-je): the science dealing with the study of cells.

cytoplasm (si´to-plazm´): the protoplasm of a cell located outside of the nucleus.

dendrite (den´drite): a nerve cell process that transmits impulses toward a neuron cell body.

dentin (den´tine): the principal substance of a tooth, covered by enamel over the crown and by cementum on the root.

dermis (der´mis): the second, or deep, layer of skin beneath the epidermis.

descending colon: the segment of the large intestine that descends on the left side from the level of the spleen to the level of the left iliac crest.

diaphragm (di´ah-fram): a flat dome of muscle and connective tissue that separates the thoracic and abdominal cavities.

diaphysis (di-af´i-sis): the shaft of a long bone.

diastole (di-as´to-le): the sequence of the cardiac cycle during which the ventricular heart chamber wall is relaxed.

diarthrosis (di´ar-thro´sis): a freely movable joint.

distal (dis´tal): away from the midline or origin; the opposite of *proximal*.

dorsal (dor´sal): pertaining to the back or posterior portion of a body part; the opposite of *ventral*.

ductus deferens (duk´tus def´er-enz): a tube that carries spermatozoa from the epididymis to the ejaculatory duct: also called the *vas deferens* or *seminal duct*.

duodenum (du´o-num): the first portion of the small intestine.

dura mater (du´rah ma´ter): the outermost meninx covering the central nervous system.

eccrine gland (ek´rin): a sweat gland that functions in body cooling.

ectoderm (ek´to-derm): the outermost of the three primary embryonic germ layers.

edema (e-de´mah): an excessive retention of fluid in the body tissues.

effector (ef-fek´tor): an organ such as a gland or muscle that responds to motor stimulation.

efferent (ef´er-ent): conveying away from the center of an organ or structure.

ejaculation (e-jak´u-la´shun): the discharge of semen from the male urethra during climax.

electrocardiogram (e-lek´tro-kar´de-o-gram´): a recording of the electrical activity that accompanies the cardiac cycle; also called ECG or EKG.

electroencephalogram (e-lek´tro-en-sef´ah-lo-gram): a recording of the brain wave pattern; also called EEG.

electromyogram (e-lek´tro-mi´o-gram): a recording of the activity of a muscle during contraction: also called EMG.

electrolyte (e-lek´tro-lite): a solution that conducts electricity by means of charged ions.

electron (e-lek´tron): the unit of negative electricity.

enamel (en-am´el): the outer, dense substance covering the crown of a tooth.

endocardium (en´do-kar´de-um): the fibrous lining of the heart chambers and valves.

endochondral bone (en´do-kon´dral): bones that form as hyaline cartilage models first and then are ossified.

endocrine gland (en´do-krine): hormone producing gland that is part of the endocrine system.

endoderm (en´do-derm): the innermost of the three primary germ layers of an embryo.

endometrium (en´do-me´tre-um): the inner lining of the uterus.

endothelium (en´do-the´le-um): the layer of epithelial tissue that forms the thin inner lining of blood vessels and heart chambers.

eosinophil (e´o-sin´o-fil): a type of white blood cell that becomes stained by acidic eosin dye; constitutes about 2%–4% of the white blood cells.

epicardium (ep´i-kar´de-um): the thin, outer layer of the heart: also called the *visceral pericardium.*

epidermis (ep´i-der´mis): the outermost layer of the skin, composed of stratified squamous epithelium.

epididymis (ep´i-did´i-mis): a coiled tube located along the posterior border of the testis; stores spermatozoa and discharges them during ejaculation.

epidural space (ep´i-du´ral): a space between the spinal dura mater and the bone of the vertebral canal.

epiglottis (ep´i-glot´is): a leaflike structure positioned on top of the larynx that covers the glottis during swallowing.

epinephrine (ep´i-nef´rin): a hormone secreted from the adrenal medulla resulting in action similar to those from sympathetic nervous system stimulation; also called *adrenaline.*

epiphyseal plate (ep´i-fize-al): a cartilaginous layer located between the epiphysis and diaphysis of a long bone and functions in longitudinal bone growth.

epiphysis (e-pif´i-sis): the end segment of a long bone, distinct in early life but later becoming part of the larger bone.

epithelial tissue (ep´i-the´le-al): one of the four basic tissue types; the type of tissue that covers or lines all exposed body surfaces.

erythrocyte (e-rith´ro-site): a red blood cell.

esophagus (e-sof´ah-gus): a tubular organ of the GI tract that leads from the pharynx to the stomach.

estrogen (es´tro-jen): female sex hormone secreted from the ovarian (Graafian) follicle.

eustachian canal (u-sta´ke-an): see *auditory tube.*

excretion (eks-kre´shun): discharging waste material.

exocrine gland (ek´so-krin): a gland that secretes its product to an epithelial surface, directly or through ducts.

expiration (ek´spi-ra´shun): the process of expelling air from the lungs through breathing out; also called *exhalation.*

extension (ek-sten´shun): a movement that increases the angle between two bones of a joint.

external ear: the outer portion of the ear, consisting of the auricle (pinna), external auditory canal.

extracellular (esk-trah-sel´u-lar): outside a cell or cells.

extrinsic (eks-trin´sik): pertaining to an outside or external origin.

facet (fas´et): a small, smooth surface of a bone where articulation occurs.

fallopian tube (fal-lo´pe-an): see *uterine tube.*

fascia (fash´e-ah): a tough sheet of fibrous connective tissue binding the skin to underlying muscles or supporting and separating muscle.

fasciculus (fah-sik´u-lus): a bundle of muscle or nerve fibers.

feces (fe´sez): waste material expelled from the GI tract during defecation, composed of food residue, bacteria, and secretions; also called *stool.*

fetus (fe´tus): the unborn offspring during the last stage of prenatal development.

filtration (fil-tra´shun): the passage of a liquid through a filter or a membrane.

fimbriae (fim´bre-e): fringelike extensions from the borders of the open end of the uterine tube.

fissure (fish´ure): a groove or narrow cleft that separates two parts of an organ.

flexion (flek´shun): a movement that decreases the angle between two bones of a joint; opposite of extension.

fontanel (fon´tah-nel): a membranous-covered region on the skull of a fetus or baby where ossification has not yet occurred: also called a *soft spot.*

foot: the terminal portion of the lower extremity, consisting of the tarsus, metatarsus, and digits.

foramen (fo-ra´men): an opening in an anatomical structure for the passage of a blood vessel or a nerve.

foramen ovale (o-val´e): the opening through the interatrial septum of the fetal heart.

fossa (fos´ah): a depressed area, usually on a bone.

fourth ventricle (ven´tri-k´l): a cavity within the brain containing cerebrospinal fluid.

fovea centralis (fo´ve-ah sen´tra´lis): a depression on the macula lutea of the eye where only cones are located, which is the area of keenest vision.

gallbladder: a pouchlike organ, attached to the inferior side of the liver, which stores and concentrates bile.

gamete (gam´ete): a haploid sex cell, sperm or egg.

gamma globulins (gam´mah glob´u-lins): protein substances often found in immune serums that act as antibodies.

ganglion (gang´gle-on): an aggregation of nerve cell bodies outside the central nervous system.

gastrointestinal tract (gas´tro-in-tes´tin-al): the tubular portion of the digestive system that includes the stomach and the small and large intestines; also called the *GI tract.*

gene (jene): one of the biologic units of heredity; parts of the DNA molecule located in a definite position on a certain chromosome.

genetics (je-net´iks): the study of heredity.

gingiva (jin-ji´vah): the fleshy covering over the mandible and maxilla through which the teeth protrude within the mouth; also called the *gum.*

gland: an organ that produces a specific substance or secretion.

glans penis (glanz pe´nis): the enlarged, distal end of the penis.

glomerular capsule (glo-mer´u-lar): the double-walled proximal portion of a renal tubule that encloses the glomerulus of a *nephron;* also called *Bowman's capsule.*

glomerulus (glo-mer´u´lus): a coiled tuft of capillaries that is surrounded by the glomerular capsule and filters urine from the blood.

glottis (glot´is): a slitlike opening into the larynx, positioned between the vocal folds.

glycogen (gli´ko-jen): the principal storage carbohydrate in animals. It is stored primarily in the liver and is made available as glucose when needed by the body cells.

goblet cell: a unicellular gland within columnar epithelia that secretes mucus.

gonad (go´nad): a reproductive organ, testis or ovary, that produces gametes and sex hormones.

gray matter: the portion of the central nervous system that is composed of nonmyelinated nervous tissue.

greater omentum (o-men´tum): a double-layered peritoneal membrane that originates on the greater curvature of the stomach and extends over the abdominal viscera.

gut: pertaining to the intestine; generally a developmental term.

gyrus (ji´rus): a convoluted elevation or ridge.

hair: an epidermal structure consisting of keratinized dead cells that have been pushed up from a dividing basal layer.

hair cells: specialized receptor nerve endings for responding to sensations, such as in the spiral organ of the inner ear.

hair follicle (fol´li-k´l): a tubular depression in the skin in which a hair develops.

hand: the terminal portion of the upper extremity, consisting of the carpus, metacarpus, and digits.

hard palate (pal´at): the bony partition between the oral and nasal cavities, formed by the maxillae and palatine bones.

haustra (haws´trh): sacculations or pouches of the colon.

haversian system (ha-ver´shan): see *osteon.*

heart: a muscular, pumping organ positioned in the thoracic cavity.

hematocrit (he-mat´o-krit): the volume percentage of red blood cells in whole blood.

hemoglobin (he´mo-glo´bin): the pigment of red blood cells that transports O_2 and CO_2.

hemopoiesis (hem´ah-poi-e´sis): production of red blood cells.

hepatic portal circulation (por´tal): the return of venous blood from the digestive organs and spleen through a capillary network within the liver before draining into the heart.

heredity (re-red´i-te): the transmission of certain characteristics, or traits, from parents to offspring, via the genes.

hiatus (hi-a´tus): an opening or fissure.

hilum (hi´lum): a concave or depressed area where vessels or nerves enter or exit an organ.

histology (his-tol´o-je): microscopic anatomy of the structure and function of tissues.

homeostasis (ho-me-o-sta´sis): a consistency and uniformity of the internal body environment which maintains normal body function.

hormone (hor´mone): a chemical substance that is produced in an endocrine gland and secreted into the bloodstream to cause an effect in a specific target organ.

hyaline cartilage (hi´ah-line): the most common kind of cartilage in the body, occurring at the articular ends of bones, in the trachea, and within the nose, and forms the precursor to most of the bones of the skeleton.

hymen (hi´men): a developmental remnant (vestige) of membranous tissue that partially covers the vaginal opening.

hyperextension (hi´per-ek-sten´shun): extension beyond the normal anatomical position of 180°.

hypothalamus (hi´po-thal´ah-mus): a structure within the brain below the thalamus, which functions as an autonomic center and regulates the pituitary gland.

ileocecal valve (il´e-o-se´kal): a specialization of the mucosa at the junction of the small and large intestine that forms a one-way passage and prevents the backflow of food materials.

ileum (il´e-um): the terminal portion of the small intestine between the jejunum and cecum.

inferior vena cava (ve´nah ka´vah): a systemic vein that collects blood from the body regions inferior to the level of the heart and returns it to the right atrium.

inguinal (ing´gwi-nal): pertaining to the groin region.

inguinal canal: the passage in the abdominal wall through which a testis descends into the scrotum.

insertion: the more movable attachment of a muscle, usually more distal in location.

inspiration (in´spi-ra´shun): the act of breathing air into the alveoli of the lungs; also called *inhalation.*

integument (in-teg´u-ment): pertaining to the skin.

internal ear: the innermost portion or chamber of the ear, containing the cochlea and the vestibular organs.

interstitial (in-ter-stish´al): pertaining to spaces or structures between the functioning active tissue of any organ.

intracellular (in-trah-sel´u-lar): within the cell itself.

intervertebral disc (in´ter-ver´te-bral): a pad of fibrocartilage between the bodies of adjacent vertebrae.

intestinal gland (in-tes´ti-nal): a simple tubular digestive gland that opens onto the surface of the intestinal mucosa and secretes digestive enzymes; also called *crypt of Lieberkuhn.*

intrinsic (in-trin´sik): situated in or pertaining to an internal organ.

iris (i´ris): the pigmented, vascular tunic portion of the eye that surounds the pupil and regulates its diameter.

islets of Langerhans (i´lets of lahng´er-hanz): see *pancreatic islets.*

isotope (i´so-tope): a chemical element that has the same atomic number as another but a different atomic weight.

isthmus (is´mus): a narrow neck or portion of tissue connecting two structures.

jejunum (je-joo´num): the middle portion of the small intestine, located between the duodenum and the ileum.

joint capsule (kap´sule): the fibrous tissue that encloses the joint cavity of a synovial joint.

jugular (jug´u-lar): pertaining to the veins of the neck which drain the areas supplied by the carotid arteries.

karyotype (kar´e-o-tip): the arrangement of chromosomes that is characteristic of the species or of a certain individual.

keratin (ker´ah-tin): an insoluble protein present in the epidermis and in epidermal derivatives such as hair and nails.

kidney (kid´ne): one of the paired organs of the urinary system that contains nephrons and filters wastes from the blood in the formation of urine.

labia majora (la´be-ah ma-jor´ah): a portion of the external genitalia of a female, consisting of two longitudinal folds of skin extending downward and backward from the mons pubis.

labia minora (ma-nor´ah): two small folds of skin, devoid of hair and sweat glands, lying between the labia majora of the external genitalia of a female.

lacrimal gland (lak´ri-mal): a tear-secreting gland, located on the superior lateral portion of the eyeball underneath the upper eyelid.

lactation (lak-ta´shun): the production and secretion of milk by the mammary glands.

lacteal (lak´te-al): a small lymphatic duct within a villus of the small intestine.

lacuna (lah-ku´nah): a hollow chamber that houses an osteocyte in mature bone tissue or a chondrocyte in cartilage tissue.

lamella (lah-mel´ah): a concentric ring of matrix surrounding the central canal in an osteon of mature bone tissue.

large intestine: the last major portion of the GI tract, consisting of the cecum, colon, rectum, and anal canal.

larynx (lar´inks): the structure located between the pharynx and trachea that houses the vocal folds (cords); commonly called the *voice box.*

lens (lenz): a transparent refractive structure of the eye, derived from ectoderm and positioned posterior to the pupil and iris.

leukocyte (lu´ko-site): a white blood cell; also spelled *leucocyte.*

ligament (lig´ah-ment): a fibrous band or cord of connective tissue that binds bone to bone to strengthen and provide support to the joint; also may support viscera.

limbic system (lim´bik): a portion of the brain concerned with emotions and autonomic activity.

linea alba (lin´e-ah al´bah): a fibrous band extending down the anterior medial portion of the abdominal wall.

locus (lo´kus): the specific location or site of a gene within the chromosome.

lumbar (lum´bar): pertaining to the region of the loins.

lumen (lu´men): the space within a tubular structure through which a substance passes.

lung: one of the two major organs of respiration within the thoracic cavity.

lymph (limf): a clear fluid that flows through lymphatic vessels.

lymph node: a small, ovoid mass located along the course of lymph vessels.

lymphocyte (lim´fo-site): a type of white blood cell characterized by a granular cytoplasm.

macula lutea (mak´u-lah lu´te-ah): a depression in the retina that contains the fovea centralis, the area of keenest vision.

malignant (mah-lig´nant): a disorder that becomes worse and eventually causes death, as in cancer.

malnutrition (mal-nu-trish´un): any abnormal assimilation of food; receiving insufficient nutrients.

mammary gland (mam´er-e): the gland of the female breast responsible for lactation and nourishment of the young.

marrow (mar´o): the soft vascular tissue that occupies the inner cavity of certain bones and produces blood cells.

matrix (ma´triks): the intercellular substance of a tissue.

meatus (me-a´tus): an opening or passageway into a structure.

mediastinum (me´de-as-ti´num): the partition in the center of the thorax between the two pleural cavities.

medulla (me-dul´ah): the center portion of an organ.

medulla oblongata (ob´long-ga´tah): a portion of the brain stem between the pons and the spinal cord.

medullary cavity (med´u-lar´e): the hollow center of the diaphysis of a long bone, occupied by marrow.

meiosis (mi-o´sis): cell division by which gametes, or haploid sex cells, are formed.

melanocyte (mel´ah-no-site): a pigment-producing cell in the deepest epidermal layer of the skin.

membranous bone (mem´brah-nus): bone that forms from membranous connective tissue rather than from cartilage.

menarche (me-nar´ke): the first menstrual discharge.

meninges (me-nin´jez): a group of three fibrous membranes that cover the central nervous system.

menisci (men-is´si): wedge-shaped cartilages in certain synovial joints.

menopause (men´o-pawz): the cessation of menstrual periods in the human female.

menses (men´sez): the monthly flow of blood from the female genital tract.

menstrual cycle (men´stru-al): the rhythmic female reproductive cycle, characterized by changes in hormone levels and physical changes in the uterine lining.

menstruation (men´stru-a´shun): the discharge of blood and tissue from the uterus at the end of the menstrual cycle.

mesentery (mes´en-ter´e): a fold of peritoneal membrane that attaches an abdominal organ to the abdominal wall.

mesoderm (mes´o-derm): the middle one of the three primary germ layers.

mesothelium (mes´o-the´leum): a simple squamous epithelial tissue that lines body cavities and covers visceral organs; also called *serosa*.

metabolism (me-tab´o-lizm): the chemical changes that occur within a cell.

metacarpus (met´ah-kar´pus): the region of the hand between the wrist and the digits, including the five bones that support the palm of the hand.

metastasis (me-tas´tah-sis): the spread of a disease from one organ or body part to another.

metatarsus (met´ah-tar´sus): the region of the foot between the ankle and the digits, containing five bones.

microbiology (mi-kro-bi-ol´o-je): the science dealing with microscopic organisms, including bacteria, fungi, viruses, and protozoa.

microvilli (mi´kro-vil´i): microscopic, hairlike projections of cell membranes on certain epithelial cells.

midbrain: the portion of the brain between the pons and the forebrain.

middle ear: the middle of the three ear chambers, containing the three auditory ossicles.

mitosis (mi-to´sis): the process of cell division, in which the two daughter cells are identical and contain the same number of chromosomes.

mitral valve (mi´tral): the left atrioventricular heart valve; also called the bicuspid valve.

mixed nerve: a nerve containing both motor and sensory nerve fibers.

molecule (mol´e-kule): a minute mass of matter, composed of a combination of atoms that form a given chemical substance or compound.

motor neuron (nu´ron): a nerve cell that conducts action potential away from the central nervous system and innervates effector organs (muscles and glands); also called *efferent neuron*.

motor unit: a single motor neuron and the muscle fibers it innervates.

mucosa (mu-ko´sah): a mucous membrane that lines cavities and tracts opening to the exterior.

muscle (mus´el): an organ adapted to contract; three types of muscle tissue are cardiac, smooth, and skeletal.

mutation (mu-ta´shun): a variation in an inheritable characteristic, a permanent transmissible change in which the offspring differ from the parents.

myelin (mi´e-lin): a lipoprotein material that forms a sheathlike covering around nerve fibers.

myocardium (mi´o-kar´de-um): the cardiac muscle layer of the heart.

myofibril (mi´o-fi´bril): a bundle of contractile fibers within muscle cells.

myoneural junction (mi´o-nu´ral): the site of contact between an axon of a motor neuron and a muscle fiber.

myosin (mi´o-sin): a thick filament protein that together with actin causes muscle contraction.

nail: a hardenend, keratinized plate that develops from the epidermis and forms a protective covering on the dorsal surfaces of the digits.

nares (na´rez): the opening into the nasal cavity; also called *nostrils*.

nasal cavity (na´zal): a mucosa-lined space above the oral cavity, which is divided by a nasal septum and is the first chamber of the respiratory system.

nasal septum (sep´tum): a bony and cartilaginous partition that separates the nasal cavity into two portions.

nephron (nef´ron): the functional unit of the kidney, consisting of a glomerulus, glomerular capsule, convoluted tubules, and the loop of the nephron.

nerve: a bundle of nerve fibers outside the central nervous system.

neurofibril node (nu´ro-fi´bril): a gap in the myelin sheath of a nerve fiber; also called the *node of Ranvier*.

neuroglia (nu-rog´le-ah): specialized supportive cells of the central nervous system.

neurolemmocyte (nu´ri-lem-o´site): a specialized neuroglia cell that surrounds an axon fiber of a peripheral neuron and forms the neurilemmal sheath; also called the *Schwann cell*.

neuron (nu´ron): the structural and functional unit of the nervous system, composed of a cell body, dendrites, and an axon; also called a *nerve cell.*

neutrophil (nu´tro-fil): a type of phagocytic white blood cell.

nipple: a dark pigmented, rounded projection at the tip of the breast.

node of Ranvier (rah-ve-a´): see *neurofibril node.*

notochord (no´to-kord): a flexible rod of tissue that extends the length of the back of an embryo.

nucleus (nu´kle-us): a spheroid body within a cell that contains the genetic factors of the cell.

nurse cells: specialized cells within the testes that supply nutrients to developing spermatozoa; also called *sertoli cells* or *sustentacular cells.*

olfactory (ol-fak´to-re): pertaining to the sense of smell.

oocyte (o´o-site): a developing egg cell.

oogenesis (o´o-hen´e-sis): the process of female gamete formation.

optic (op´tik): pertaining to the eye and the sense of vision.

optic chiasma (ki-as´mah): an X-shaped structure on the inferior aspect of the brain where there is a partial crossing over of fibers in the optic nerves.

optic disc: a small region of the retina where the fibers of the ganglion neurons exit from the eyeball to form the optic nerve; also called the *blind spot.*

oral: pertaining to the mouth; also called *buccal.*

organ: a structure consisting of two or more tissues, which performs a specific function.

organelle (or´gan-el´): a minute structure of a cell with a specific function.

organism (or´gah-nizm): an individual living creature.

orifice (or´i fis): an opening into a body cavity or tube.

origin (or´i-jin): the place of muscle attachment onto the more stationary point or proximal bone; opposite the insertion.

osmosis (os-mo´sis): the passage of a solvent, such as water, from a solution of lesser concentration to one of greater concentration through a semipermeable membrane.

ossicle (os´si-l´l): one of the three bones of the middle ear.

osteocyte (os´te-o-site): a mature bone cell.

osteon (os´te-on): a group of osteocytes and concentric lamellae surrounding a central canal within bone tissue; also called a *haversian system.*

ovarian follicle (o-va´re-an fol´li-k´l): a developing ovum and its surrounding epithelial cells.

ovary (o´vah-re): the female gonad in which ova and certain sexual hormones are produced.

oviduct (o´vi-dukt): the tube that transports ova from the ovary to the uterus; also called the *uterine tube* or *fallopian tube.*

ovulation (o´vu-la´shun): the rupture of an ovarian follicle with the release of an ovum.

ovum (o´vum): a secondary oocyte after ovulation but before fertilization.

palate (pal´at): the roof of the oral cavity.

palmar (pal´mar): pertaining to the palm of the hand.

pancreas (pah´kre-as): organ in the abdominal cavity that secretes gastric juices into the GI tract and insulin and glucagon into the blood.

pancreatic islets (pan´kre-at´ik): a cluster of cells within the pancreas that forms the endocrine portion of the pancreas; also called *islets of Langerhans.*

papillae (pah-pil´e): small nipplelike projections.

paranasal sinus (par´ah-na´zal si´nus): an air chamber lined with a mucous membrane that communicates with the nasal cavity.

parasympathetic (par´ah-smi´pah-thet´ik): pertaining to the division of the autonomic nervous system concerned with activities that are antagonistic to sympathetic.

parathyroids (par´ah-thi´roids): small endocrine glands that are embedded on the posterior surface of the thyroid glands and are concerned with calcium metabolism.

parietal (pah-ri´e-tal): pertaining to a wall of an organ or cavity.

parotid gland: (pah-rot´id): one of the paired salivary glands on the side of the face over the masseter muscle.

parturition (par´tu-rish´un): the process of childbirth.

pathogen (path´o-jen): any disease-producing organism.

pectoral girdle (pek´to-ral): the portion of the skeleton that supports the upper extremities.

pelvic (pel´vik): pertaining to the pelvis.

pelvic girdle: the portion of the skeleton to which the lower extremities are attached.

penis (pe´nis): the external male genital organ, through which urine passes during urination and which transports semen to the female during coitus.

pericardium (per´i-kar´de-um): a protective serous membrane that surrounds the heart.

perineum (per´i-ne´um): the floor of the pelvis, which is the region between the anus and the scrotum in the male and between the anus and the vulva in the female.

periosteum (per´e-os´te-um): a fibrous connective tissue covering the outer surface of bone.

peripheral nervous system (pe-rif´er-al): the nerves and ganglia of the nervous system that lie outside of the brain and spinal cord.

peristalsis (per´i-stal´sis): rhythmic contractions of smooth muscle in the walls of various tubular organs, which move the contents along.

peritoneum (per´i-to-ne´um): the serous membrane that lines the abdominal cavity and covers the abdominal viscera.

phagocyte (fag´o-site): any cell that engulfs other cells, including bacteria, or small foreign particles.

phalanx (fa´lanks), pl. *phalanges:* a bone of the finger or toe.

pharynx (far´inks): the organ of the GI tract and respiratory system located at the back of the oral and nasal cavities and extending to the larynx anteriorly and the esophagus posteriorly; also called the *throat.*

physiology (fiz´e-ol´o-je): the science that deals with the study of body functions.

pia mater (pi´ah ma´ter): the innermost meninx that is in direct contact with the brain and spinal cord.

pineal gland (pin´e-al): a small cone-shaped gland located in the roof of the third ventricle.

pituitary gland (pi-tu´i-tar´e): a small, pea-shaped endocrine gland situated on the inferior surface of the brain that secretes a number of hormones; also called the *hypophysis* and commonly called the *"master gland."*

placenta (plah-sen´tah): the organ of metabolic exchange between the mother and the fetus.

plasma (plaz´mah): the fluid, extracellular portion of circulating blood.

platelets (plate´lets): fragments of specific bone marrow cells that function in blood coagulation: also called *thrombocytes.*

pleural membranes (ploor´al): serous membranes that surround the lungs and line the thoracic cavity.

plexus (plek´sus): a network of enterlaced nerves or vessels.

plica circulares (pli´kah ser-ku-lar´is): a deep fold within the wall of the small intestine that increases the absorptive surface area.

pons (ponz): the portion of the brain stem just above the medulla oblongata and anterior to the cerebellum.

posterior (dorsal) (pos-ter´e-or): toward the back.

pregnancy: a condition where a female has a developing offspring in the uterus.

prenatal (pre-na´tal): the period of offspring development during pregnancy; before birth.

proprioceptor (pro′pre-o-sep′tor): a sensory nerve ending that responds to changes in tension in a muscle or tendon.

prostate (pros′tate): a walnut-shaped gland surrounding the male urethra just below the urinary bladder that secretes an additive to seminal fluid during ejaculation.

proximal (prok′si-mal): closer to the midline of the body or origin of an appendage: opposite of *distal.*

puberty (pu′ber-te): the period of development in which the reproductive organs become functional.

pulmonary (pul′mo-ner′e): pertaining to the lungs.

pupil: the opening through the iris that permits light to enter the vitreous chamber of the eyeball and be refracted by the lens.

receptor (re-sep′tor): a sense organ or a specialized end of a sensory neuron that receives stimuli from the environment.

rectum (rek′tum): the portion of the GI tract between the sigmoid colon and the anal canal.

reflex arc: the basic conduction pathway through the nervous system, consisting of a sensory neuron, interneuron, and a motor neuron.

renal (re′nal): pertaining to the kidney.

renal corpuscle (kor′pus′l): the portion of the nephron consisting of the glomerulus and a glomerular capsule.

renal pelvis: the inner cavity of the kidney formed by the expanded ureter and into which the calyces open.

replication (re-pli-ka′shun): the process of producing a duplicate; a copying or duplication, such as DNA replication.

respiration: (res′pi-ra′shun): the exchange of gases between the external environment and the cells of an organism.

rete testis (re′te tes′tis): a network of ducts in the center of the testis, site of spermatozoa production.

retina (ret′i-nah): the inner layer of the eye that contains the rods and cones.

retraction (re-trak′shun): the movement of a body part, such as the mandible, backward on a plane parallel with the ground; the opposite of *protraction.*

rod: a photoreceptor in the retina of the eye that is specialized for colorless, dim light vision.

rotation (ro-ta′shun): the movement of a bone around its own longitudinal axis.

rugae (ru′je): the folds or ridges of the mucosa of an organ.

sagittal (saj′i-tal): a vertical plane through the body that divides it into right and left portions.

salivary gland (sal′i-ver-e): an accessory digestive gland that secretes saliva into the oral cavity.

sarcolemma (sar′ko-lem′ah): the cell membrane of a muscle fiber.

sarcomere (sar′ko-mere): the portion of a skeletal muscle fiber between the two adjacent Z lines that is considered the functional unit of a myofibril.

Schwann cell (shwahn): see *neurolemmocyte.*

sclera (skle′rah): the outer white layer of connective tissue that forms the protective covering of the eye.

scrotum (skro′tum): a pouch of skin that contains the testes and their accessory organs.

sebaceous gland (se-ba′shus): an exocrine gland of the skin that secretes *sebum,* an oily protective product.

semen (se′men): the secretion of the reproductive organs of the male, consisting of spermatozoa and additives.

semicircular ducts: tubular channels within the inner ear that contain the receptors for equilibrium; also called *semicircular canals.*

semilunar valve (sem′e-lu′nar): crescent-shaped heart valves, positioned at the entrances to the aorta and the pulmonary trunk.

seminal vesicles (sem′i-nal ves′i-k′lz): a pair of accessory male reproductive organs lying posterior and inferior to the urinary bladder, which secrete additives to spermatozoa into the ejaculatory ducts.

sensory neuron (nu′ron): a nerve cell that conducts an impulse from a receptor organ to the central nervous system; also called *afferent neuron.*

serous membrane (se′rus): an epithelial and connective tissue membrane that lines body cavities and covers viscera; also called *serosa.*

sesamoid bone (ses′ah-moid): a membranous bone formed in a tendon in response to joint stress.

sigmoid colon (sig′moid ko′lon): the S-shaped portion of the large intestine between the descending colon and the rectum.

sinoatrial node (sin′o-a′tre-al): a mass of cardiac tissue in the wall of the right atrium that initiates the cardiac cycle; the SA node; also called the *pacemaker.*

sinus (si′nus): a cavity or hollow space within a body organ such as a bone.

skeletal muscle: a type of muscle tissue that is multinucleated, occurs in bundles, has crossbands of proteins, and contracts either in a voluntary or involuntary fashion.

small intestine: the portion of the GI tract between the stomach and the cecum, functions in absorption of food nutrients.

smooth muscle: a type of muscle tissue that is nonstriated, composed of fusiform, single-nucleated fibers, and contracts in an involuntary, rhythmic fashion within the walls of visceral organs.

somatic (so-mat′ik): pertaining to the nonvisceral parts of the body.

spermatic cord (sper′mat′ik): the structure of the male reproductive system composed of the ductus deferens, spermatic vessels, nerve, cremasteric muscle, and connective tissue.

spermatogenesis (sper′mah-to-jen′e-sis): the production of male sex gametes, or spermatozoa.

spermatozoon (sper′mah-to-zo′on): a sperm cell, or gamete.

sphincter (sfingk′ter): a circular muscle that constricts a body opening or the lumen of a tubular structure.

spinal cord (spi′nal): the portion of the central nervous system that extends from the brain stem through the vertebral canal.

spinal nerve: one of the thirty-one pairs of nerves that arise from the spinal cord.

spleen: a large, blood-filled organ located in the upper left of the abdomen and attached by the mesenteries to the stomach.

spongy bone (spun′je): a type of bone that contains many porous spaces; also called *cancellous bone.*

stomach: a pouchlike digestive organ between the esophagus and the duodenum.

submucosa (sub′mu-ko′sah): a layer of supportive connective tissue that underlies a mucous membrane.

superior vena cava (ve′nah ka′vah): a large systemic vein that collects blood from regions of the body superior to the heart and returns it to the right atrium.

surfactant (ser-fak′tant): a substance produced by the lungs that decreases the surface tension within the alveoli.

suture (su′chur): a type of fibrous joint articulating between bones of the skull.

sympathetic (sim′pah-thet′ik): pertaining to that part of the autonomic nervous system concerned with activities antagonistic to the parasympathetic.

synapse (sin′aps): a minute space between the axon terminal of a presynaptic neuron and a dendrite of a postsynaptic neuron.

synovial cavity (si-no´ve-al): a space between the two bones of a synovial joint, filled with synovial fluid.

system: a group of body organs that function together.

systole (sis´to-le): the muscular contraction of the ventricles of the heart during the cardiac cycle.

systolic pressure (sis´tol´ik): arterial blood pressure during the ventricular systolic phase of the cardiac cycle.

target organ: the specific body organ that a particular hormone affects.

tarsus (tahr´sus): pertaining to the ankle; the proximal portion of the foot that contains the seven tarsal bones.

tendo calcaneous (ten´do kal-ka´ne-us): the tendon that attaches the calf muscles to the calcaneous bone.

tendon (ten´dun): a band of dense regular connective tissue that attaches muscle to bone.

testis (tes´tis): the primary reproductive organ of a male, which produces spermatozoa and male sex hormones.

thoracic (tho-ras´ik): pertaining to the chest region.

thoracic duct: the major lymphatic vessel of the body, which drains lymph from the entire body except the upper right quadrant and returns it to the left subclavian vein.

thorax (tho´raks): the chest.

thymus gland (thi´mus): a bi-lobed lymphoid organ positioned in the upper mediastinum, posterior to the sternum and between the lungs.

tissue: an aggregation of similar cells and their binding intercellular substance, joined to perform a specific function.

tongue: a protrusible muscular organ on the floor of the oral cavity.

trachea (tra´ke-ah): the airway leading from the larynx to the bronchi; also called the *windpipe*.

tract: a bundle of nerve fibers within the central nervous system.

transverse colon (ko´lon): a portion of the large intestine that extends from right to left across the abdomen between the hepatic and splenic flexures.

tricuspid valve (tri-kus´pid): the heart valve between the right atrium and the right ventricle.

tympanic membrane (tim-pan´ik): the membranous eardrum positioned between the outer and middle ear; also called the *tympanum*, or the *ear drum*.

umbilical cord (um-bil´i-kal): a cordlike structure containing the umbilical arteries and vein, which connects the fetus with the placenta.

umbilicus (um-bil´i-kus): the site where the umbilical cord was attached to the fetus: also called the *navel*.

ureter (u-re´ter): a tube that transports urine from the kidney to the urinary bladder.

urethra (u-re´thrah): a tube that transports urine from the urinary bladder to the outside of the body.

urinary bladder (u´re-ner´e): a distensible sac in the pelvic cavity which stores urine.

uterine tube (u´ter-in): the tube through which the ovum is transported to the uterus and where fertilization takes place: also called the *oviduct* or *fallopian tube*.

uterus (u´ter-us): a hollow, muscular organ in which a fetus develops. It is located within the female pelvis between the urinary bladder and the rectum.

uvula (u´vu-lah): a fleshy, pendulous portion of the soft palate that blocks the nasopharynx during swallowing.

vagina (vah-ji´nah): a tubular organ that leads from the uterus to the vestibule of the female reproductive tract and receives the male penis during coitus.

vein: a blood vessel that conveys blood toward the heart.

ventral (ven´tral): toward the front surface of the body: also called *anterior*.

vestibular folds: the supporting folds of tissue for the vocal folds within the larynx.

vestibular window: a membrane-covered opening in the bony wall between the middle and inner ear, into which the footplate of the stapes fits; also called *oval window*.

viscera (vis´er-ah): the organs within the abdominal or thoracic cavities.

vitreous humor (vit´re-us hu´mor): the transparent gell that occupies the space between the lens and retina of the eye.

vocal folds: folds of the mucous membrane in the larynx that produce sound as they are pulled taut and vibrated; also called *vocal cords*.

vulva (vul´vah): the external genitalia of the female that surround the opening of the vagina; also called the *pudendum*.

zygote (zi´gote): a fertilized egg cell formed by the union of a sperm and an ovum.

Index